T0341328

High-Temperature Superconducting Devices for Energy Applications

High-Temperature Superconducting Devices for Energy Applications

Edited by
Raja Sekhar Dondapati

CRC Press

Taylor & Francis Group

Boca Raton London New York

CRC Press is an imprint of the
Taylor & Francis Group, an **informa** business

First edition published 2021
by CRC Press
6000 Broken Sound Parkway NW, Suite 300, Boca Raton, FL 33487-2742

and by CRC Press
2 Park Square, Milton Park, Abingdon, Oxon, OX14 4RN

© 2021 Taylor & Francis Group, LLC

CRC Press is an imprint of Taylor & Francis Group, LLC

Library of Congress Cataloging-in-Publication Data

Names: Dondapati, Raja Sekhar, editor.
Title: High-temperature superconducting devices for energy applications / edited by Raja Sekhar Dondapati.
Description: First edition. | Boca Raton, FL : CRC Press, 2021. | Includes index.
Identifiers: LCCN 2020020239 (print) | LCCN 2020020240 (ebook) | ISBN 9780367492502 (hardback) | ISBN 9781003045304 (ebook)
Subjects: LCSH: High temperature superconductors. | Electric power systems--Materials.
Classification: LCC TK454.4.S93 H49 2021 (print) | LCC TK454.4.S93 (ebook) | DDC 621.31028/4--dc23
LC record available at https://lccn.loc.gov/2020020239
LC ebook record available at https://lccn.loc.gov/2020020240

ISBN: 978-0-367-49250-2 (hbk)
ISBN: 978-1-003-04530-4 (ebk)

Typeset in Palatino
by Deanta Global Publishing Services, Chennai, India

Contents

Preface

High-Temperature Superconducting Devices for Energy Applications is unique in addressing technological issues in various engineering applications. It would not have been possible to bring this book to the desks of scientists, researchers, and engineers without the contributions of Sudheer Thadela, Gaurav Vyas, Rajesh Kumar Gadekula, and Rahul Agarwal. In addition, the continuous guidance by the publishing staff to elevate the standard of the book in terms of suggestions for enhancing the quality of illustrations is appreciated. The funding support provided by the Central Power Research Institute (CPRI), Bangalore, India to perform the feasibility study on superconducting magnetic energy storage devices, through the Grant ID: CPRI/RSOP/2019/TR/06, under the scheme "Research Scheme on Power (RSOP)", is gratefully acknowledged. The encouragement from family members during times of working on editing and authoring of this book is also appreciated and acknowledged.

Chapter 1 deals with the discovery of superconductors, which can address issues such as global energy demand and efficient energy transmission. Several macroscopic characteristics, such as zero-resistance and Meissner effect, pertaining to superconductors, are discussed. In addition, this chapter provides an overview of how a superconductor forms the foundation for superconducting technologies such as superconducting cables, generators/motors, transformers, fault current limiters, and energy storage. Chapter 2 provides an overview of the methodology used for cooling superconducting devices. The fundamentals of a cryogenic cooling system, along with various associated heat sources, are mentioned. Moreover, the various cooling strategies, such as bath cooling and forced-flow cooling and dry/cryogen-free systems pertaining to superconducting technologies for power application, are illustratively discussed. In Chapter 3, motors/generators based on superconductors are discussed. First, introduction to superconducting motors is discussed, along with the recent specifications achieved. Further, the types of superconducting motors/generators are presented, followed by the involved design analysis. Finally, the chapter concludes with the future trends in the development of rotating machinery. Chapter 4 deals with direct electrical energy storage systems and the technology for the development of superconducting magnetic energy storage (SMES) systems which has attracted researchers due to its high-power density, ultra-fast response, and high efficiency in energy conversion. This SMES is potentially suitable for short discharge time and high power applications. A detailed description on the construction and the working of SMES is presented in Chapter 4. Further, different configurations of SMES are addressed, which will depend on the challenges faced during grid integration and comparison with other storage technologies. Further, worldwide projects on SMES are also mentioned. Moreover, the superconducting wires and tapes used for the construction of superconducting magnets are described. In addition, the future application of the SMES in the electrical power grid is explained in detail. Chapter 5 discusses the technology of electric power transmission with superconducting cables. These cables have the capacity to transfer more power under the same applied voltage with reduced electrical resistance. Such a feature has attracted researchers from various scientific communities to address the challenges in the construction of high-temperature superconductors (HTS) cables. Moreover, several worldwide projects based on superconducting cables are also mentioned. In addition, configurations of HTS cables, along with design analysis, are also described. Finally, existing challenges and opportunities that superconducting cables offer for researchers to make contributions in the field of applied superconductivity are also presented. Chapter 6 throws some light on the abnormal operation of the electrical systems due to faults during operation. Faults cause imbalance of the phases, over current, under voltage, high voltage sags, and reversed power than the normal rated voltages and currents. In order to protect the devices from such faults, superconducting fault current limiters (SFCL) are used. Various designs of such SFCL are discussed in Chapter 6.

Enormous amounts of effort have been invested in addressing most of the technological issues with high-temperature superconductors in this book. However, the book is limited to energy applications only. Electronic applications of high-temperature superconductors are yet to be explored to a greater extent, which could bring about a paradigm shift in digital and information technologies.

Dr. Raja Sekhar Dondapati
School of Mechanical Engineering
Lovely Professional University
Punjab, India

Editor

Dr. Raja Sekhar Dondapati is an associate professor and research coordinator in the School of Mechanical Engineering, Lovely Professional University, Punjab. He received his PhD from IIT Kharagpur in the field of cryogenics, with reference to applied superconductivity. He contributed to the development of nuclear fusion technologies. He is a life member of various international and national societies such as International Association of Engineers (IAENG), Indian Society for Heat and Mass Transfer (ISHMT), and Indian Society for Technical Education (ISTE). He has published a number of technical papers in prestigious journals including *Cryogenics-UK, Fusion Engineering and Design* and *IEEE Transactions on Applied Superconductivity,* etc. He is a reviewer of various international journals. He has contributed several book chapters in the field of computational materials science. He is also a principal investigator of a sponsored project funded by Central Power Research Institute, Bangalore on the development of superconducting magnetic energy storage devices through funding scheme RSOP with reference CPRI/RSOP/2019/TR/06.

Contributors

Rahul Agarwal
School of Mechanical Engineering
Lovely Professional University
Phawara, Punjab, India

Raja Sekhar Dondapati
School of Mechanical Engineering
Lovely Professional University
 Phawara, Punjab, India

Rajesh Kumar Gadekula
School of Mechanical Engineering
Lovely Professional University
Phawara, Punjab, India

Sudheer Thadela
Applied Materials
Bengalore, India

Gaurav Vyas
School of Mechanical Engineering
Lovely Professional University
Phawara, Punjab, India

1

Introduction to Superconducting Devices

Raja Sekhar Dondapati

CONTENTS

1.1 Introduction

In this chapter, the fundamentals of superconductors are introduced in order to form the foundation of the remaining chapters on high-temperature superconducting power equipment. The discussion further involves the introduction to types of superconductors, along with their magnetization characteristics. In the early 1900s, soon after the method of liquefaction of helium by Kamerlingh Onnes, further investigation begun concerning the electrical resistance of pure metals at low temperature. At that time, the electric resistance characteristic at low temperature was unexplored and unknown. The existing prediction ranged from the continuous linear decrease of electrical resistance toward leveling at some residual resistance value, or an increase at some point due to electron scattering mechanisms. At that time, one of the purest metals available was mercury. Kamerlingh Onnes, in 1911, measured the electrical resistance of pure mercury as a function of temperature and discovered that mercury's resistance abruptly dropped to zero below 4 K (operating temperature of liquid helium) [1, 2]. A new realization came that below 4 K, mercury enters a new state, which he called "superconductivity". Such remarkable and fascinating features had not been predicted, which still intrigues today. Since then, many materials were subsequently discovered in order to show the phenomenon of superconductivity at low temperatures. Since then, a vast range of categorical devices, which take up the majority of the field of physics, is in a continuous state of development, as illustrated in Figure 1.1.

Moreover, for many years, it was assumed that only one type of superconductor existed; however, it was later realized that two distinct types of superconductors exist. The two superconductors shared many properties; however, the most distinguishing feature is the characteristics they exhibit under an external magnetic field. The first kind of superconductor, the Type I superconductors (also known as "soft" superconductors, usually elements), loses its superconducting properties in a relatively weak magnetic field, whereas, Type II superconductors (also known as "hard" superconductors, usually alloys) can withstand very strong magnetic fields before losing superconducting properties. However, few elements are notable exceptions, which include Niobium, Vanadium, and Technetium, which are Type II superconductors. Further discussion on the properties of superconductors is presented in subsequent sections.

1.2 Theory of Superconductivity

In this section, the currently developed theories, which explain most of the superconductors' characteristics, are discussed.

FIGURE 1.1
Multidisciplinary aspects of superconducting technologies.

1.2.1 Two-Fluid Model

The two-fluid model is a phenomenological model, which is based on the following three basic assumptions [3]:

1. In the superconducting state, the carrier consists of superconducting electrons and normal electrons. The former transports current without resistance and the latter transports current with resistance. Hence, the carrier's density is composed of superconducting electron density and normal electron density.

2. In the superconducting state, the carrier density of the superconductor (n) is defined as the linear combination of normal electrons and superconducting electrons, given by Equation (1.1), where n_S and n_N refer to superconducting carrier density and normal carrier density, respectively.

$$n = n_S + n_N \tag{1.1}$$

3. In the superconductor, the normal current density (J_N) and the superconducting current density (J_S) mutually penetrate and transmit independently. Both are interchangeable according to different temperatures and magnetic fields, and finally constitute the total current density J of the superconductor, given by Equation (1.2),

$$J = J_S + J_N \tag{1.2}$$

4. The normal current density of the superconductor (J_N) is given by Equation (1.3), where, e_N and v_N represent the normal electron density and velocity, respectively. Due to the scattering of normal electrons by lattice vibrations, impurities, and defects, the resistance of the conductor is not zero. In a similar fashion, the superconducting current density (J_S) is given by Equation (1.4), where e_S and v_S represent the superconducting electron charge and velocity, respectively.

$$J_N = n_N e_N v_N \tag{1.3}$$

$$J_S = n_S e_S v_S \tag{1.4}$$

Based on classical mechanics, if the superconducting electrons with mass m_S are not scattered by lattice vibrations, impurities, and defects, they will accelerate in the presence of the electric field E and obey Newton's second law:

$$m_S \frac{dv_S}{dt} = e_S E \tag{1.5}$$

Combining equations (1.4) and (1.5), we obtain:

$$\frac{\partial J_S}{\partial t} = \frac{n_S e_S^2}{m_S} E \tag{1.6}$$

In under direct current (DC) operating condition, i.e., superconducting current density remains constant, the left-hand side of Equation (1.6) becomes zero; thus, $E = 0$. Now, considering Ohm's law, $J_S = \sigma E$, where σ (conductivity) should be infinite and thus resistivity $\rho = 0$. However, if the superconductor's current density varies with time ($E \neq 0$), then the electric field will cause a Joule loss. Therefore, zero resistance of the superconductor only occurs at steady operation (DC).

Moreover, to describe the variation of superconducting electron density with temperature T, an order parameter relating to temperature $\omega(T)$ is introduced, given by equation (1.7). When the temperature is higher than the critical temperature (T_c), the superconductor is in normal conductor state, making the superconducting electron density $n_S(T)$ zero; thus, $\omega(T) = 0$.

However, at $T = 0$ K, all electrons convert into superconducting electrons, making $\omega(T) = 1$. If the temperature T is in the range of $0 < T < T_c$, then the range of $\omega(T)$ and carrier density are $0 < \omega(T) < 1$ and $0 < n < n_S$, respectively.

$$\omega(T) = \frac{n_S}{n} \qquad (1.7)$$

1.2.2 Microscopic Theory – BCS Theory

The two-fluid model explains the superconducting phenomenon as a phenomenological model and does not provide fundamental insight on the mechanism of superconductivity. In order to provide deep understanding on superconductivity, with phenomena such as zero-resistance effect, the BCS theory establishes superconductivity from a microscopic point of view by considering Bose–Einstein condensation and interaction theory between electrons and lattice in quantum mechanics [4]. This theory can explain most of the superconducting phenomena. The mathematical treatment has been provided in the literature [5].

According to the BCS theory, a pair of electrons is coupled by the interaction between an electron and a phonon. Such an interaction is equivalent to direct interactions between two electrons, which makes such coupling, known as Cooper pairs, move toward the iron-core with a positive charge due to the Coulomb attractive force. Under such cases, the electron alters the positive charge distribution adjacent to the ion-core when it moves through the lattice, which leads to the creation of a local region with high positive charge distribution. Such a phenomenon results in attractive interaction with other adjacent electrons. To further explain superconductivity that results from Cooper pairs, we introduce the mechanism of resistance generation in a conventional conductor in brief. In conventional conductors, the movement of a single electron is affected by the inelastic scattering from the lattice, which causes part of the energy in the electron to be delivered to the lattice, which results in a macroscopic phenomenon of temperature rise and thus *Joule heating*. Such a phenomenon is the origin of resistance in a conventional conductor.

In a superconductor, Cooper pairs act as energy carriers. The movement of Cooper pairs causes an inelastic collision with the lattice, due to which it loses a part of its energy to the lattice. However, due to the opposite wave vector of the other electron in the Cooper pair it will simultaneously gain the same energy from the lattice by its inelastic collision with the lattice. As a consequence, the net energy of the Cooper pair does not change in the whole scattering process; thus, there is no energy loss. Hence, there is no resistance for the movement of the carrier, resulting in superconductivity. When the temperature reduces to absolute zero, i.e., $T = 0$ K, all the electrons convert into superconducting electrons. With further increase in temperature, electrons are generated when each Cooper pair is destroyed, which decreases the number of Cooper pairs and consequently increases the number of normal electrons. When the temperature reaches its critical transition temperature T_c, the Cooper pair disappears, which leads to the transition of a superconductor to a normal conductor, leading to its quenching.

1.3 Fundamental Properties of Superconductivity

1.3.1 Zero Resistance Characteristics

The zero resistance characteristics of a superconductor refer to the phenomenon of abrupt disappearance of resistance at a certain temperature, known as the critical temperature of a superconductor (T_c). In such conditions, the superconductor can transport direct current (DC) without resistance. Further, if a closed loop is formed in which a current is induced, the current, known as "persistent current", will flow indefinitely without any signs of decay for several years. The upper limit of resistivity measured for "persistent current" experiment is less than 10–27 Ω-m, whereas a good conventional conductor, such as copper, has a resistivity of 10–10 Ω-m at 4.2 K. Thus, the resistivity of copper is more than 17 orders of magnitude, higher than that of a superconductor [5]. A typical dependence of resistance on the temperature of a superconductor and a normal conductor is shown in Figure 1.2. The resistance of

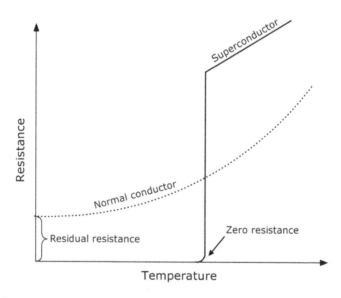

FIGURE 1.2

Comparison between resistance v/s temperature curves of a superconductor and normal conductor.

a superconductor drops to zero when the temperature reduces to a certain value below the critical temperature (T_c), which is due to the non-scattering by crystal lattice, resulting in no dissipation of energy when operating with DC. Whereas for a normal conductor, upon decreasing temperature, some resistance in the material is observed.

Ever since the discovery of superconductivity in 1911, many attempts have been made to establish and describe the physical characteristics of a superconductor. Some simple and easy to comprehend models, such as two-fluid are established, which are relatively intuitive theoretically. This model can successfully describe the motion and magnetic field distribution of carriers within the superconductor. Combined with constitutive Maxwell's electromagnetism equations, the two-fluid model provides an explanation for some superconducting phenomena, such as zero-resistance and Meissner effect. In 1957, Bardeen, Cooper, and Schrieffer, based on a series of interactions between electrons and lattice in quantum mechanics, proposed the concept of Cooper pairs and established the well-known Barden–Cooper–Schrieffer (BCS) theory, a superconducting quantum theory which describes superconductivity from a microscopic point of view. This model can successfully explain most of the superconducting phenomena.

1.3.2 Perfect Diamagnetism – Meissner Effect

In 1933, Meissner and Ochsenfeld [6] began an investigation on Type I superconductors. They discovered that on cooling a superconductor below its superconducting transition temperature, placed in an applied steady-state magnet field, the magnetic field lines were completely expelled from the interior of the superconductor. This phenomenon is called Meissner effect. Such behavior is possible under magnetization in the opposite sense ($M = -H$); a perfect diamagnetism is exhibited. Such type of magnetization measurement is sometimes referred to as the field-cooled (FC) experiment and is schematically shown in Figure 1.3 (left). However, this characteristic is far different from a zero-field-cooled (ZFC) experiment and cannot be simply explained by assuming that a superconductor is a perfect conductor (infinite mean free path). Instead, the Meissner effect implies that the flux density inside the material is zero ($B = 0$) for temperature below its critical transition temperature (T_c). If the superconductor would simply be a perfect conductor with infinite conductivity ($\sigma = \infty$) and was cooled below T_c in the presence of a steady-state magnetic field (H), there would be no magnet flux expulsion ($B = 0$) at T_c. Such a perfect conductor cooled in the background steady-state magnet field would simply pass through it. If the perfect conductor was cooled in a zero magnetic field, upon subsequently applying the magnetic field, the perfect conductor would repel flux. The whole process is illustrated in Figure 1.3. Hence, perfect diamagnetism under FC magnetization is the true signature of superconductivity.

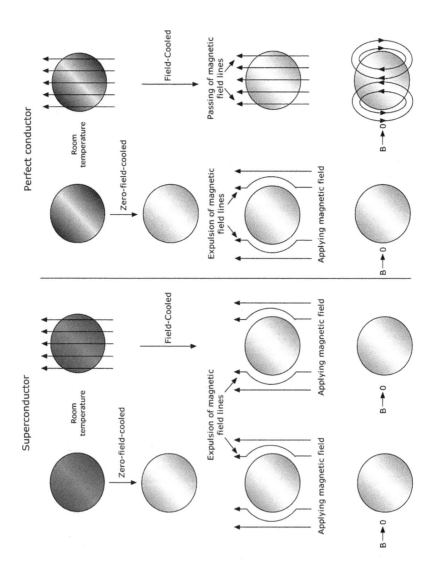

FIGURE 1.3

Characteristics of a superconductor and perfect conductor in the presence of an external magnetic field.

1.3.3 London Equations and Magnetic Penetration Depth

The London brothers in 1935 [3] were able to mathematically describe the Meissner effect by assuming that the current density J in the superconductor is directly proportional to the vector potential A of the local magnetic field B, where $B = \nabla \times A$. The London equation is given by $J = -1/(4\pi\lambda^2_L)A$ or $J = -1/(\mu_0\lambda^2_L)A$ (cgs/mks), where λ_L has the dimensions of length.

Under static conditions, one of Maxwell's equations (Ampere's law) reduces to $\nabla \times B = 4\pi J$ or $\nabla \times B = \mu_0 J$ (cgs/mks). By taking curl on both sides, this equation reduces to $-\nabla^2 \times B = 4\pi\nabla \times J$ or $-\nabla^2 B = \mu_0\nabla \times J$ (cgs/mks). By combining this equation with the curl of the London equation, we obtain $\nabla^2 B = B/\lambda^2_L$. The solution of this equation is a flux density, which decays exponentially with the distance from the external superconductor surface. Considering a one-dimensional, semi-infinite superconductor occupying the positive side of the x-axis, the solution to this equation for the magnetic flux density inside this medium would be $B(x) = B_0\exp(-x/\lambda_L)$, where $B_0 = \mu_0 H$ is the parallel field at the plane boundary.

1.4 Critical Parameters

There are three fundamental parameters, namely, critical temperature T_c, critical external magnetic field H_c, and critical current density J_c, as shown in Figure 1.4. These three are the most important parameters which define the state of superconductivity in a superconductor.

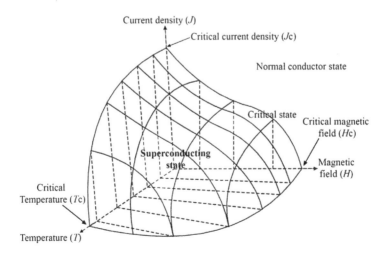

FIGURE 1.4
Critical parameters of a superconductor.

1.4.1 Critical Temperature T_c

Superconductors show superconductivity when their temperature is below a certain value, i.e., the temperature at which the superconductors transfer to a superconductor from a normal conducting state. This temperature is referred to as *critical temperature*, denoted by T_c. In metals or alloy superconductors, which have high purity, a single crystal, and is stress free, T_c is smaller than 10^{-3}. Further, the transition of practical high-temperature superconductors is usually in the range of 0.5 K to 1 K, due to the intrinsic characteristics, such as internal inhomogeneity, weak link, granularity, and defects.

1.4.2 Critical Field H_c

Superconductors lose their superconductivity when the magnetic field strength exceeds a certain value in the external magnetic field. The magnetic field strength which causes a superconductor to lose its superconductivity is called *critical field strength* and is denoted by H_c. When the temperature is below T_c, H_c is a function of temperature and continuously increases with a decrease in temperature. With a similar behavior as T_c, there is also a field transition width when the superconductors transfer from a normal state to a superconducting state. For a practical superconductor, there are usually two critical fields, namely, *lower critical field H_{c1}* and *upper critical field H_{c2}*. When the external field is less than H_{c1}, the superconductor is in Meissner state; however, when the external field is larger than H_{c2}, the superconductor is in *normal state*. While the field is between H_{c1} and H_{c2}, the superconductor is in *mixed state* (more details provided in Section 1.5).

1.4.3 Critical Current Density J_c

Although a superconductor can transport current without resistance, its ability is limited. It loses its superconductivity and converts to a normal conducting state if the transport current exceeds a certain value called the critical current I_c. In other words, I_c refers to the maximum direct current that can flow without encountering any resistive losses in the superconductor. The criterion for such conditions is the electric field strength E as 1 μV/cm or the resistivity ρ as 10–13 Ω-m. The critical current I_c continuously decreases with increase in temperature (T) and magnetic field (B).

The three fundamental critical parameters T_c, H_c, and J_c of a superconductor are dependent on each other. These parameters enclose a volume, which is constant of a superconductor. Thus, the increase of any one critical parameter will subsequently reduce the remaining two parameters and vice versa. Any point lying inside this volume will make the material a superconductor. If the point lies on the surface of J_c, T_c, and H_c, it will result in the critical state of a superconductor. Further, if the point lies outside the volume, it will result in the normal superconducting state of a conductor. Table 1.1 lists some critical parameters of several types of superconductors [7].

1.5 Classification and Magnetization

1.5.1 Coherence Length

In Section 1.3, we introduced the concept of Meissner effect and magnetic field penetration depth λ. However, for the classification of superconductors, another important microscopic parameter called *coherence length* ξ is described in the current section.

According to the BCS theory, superconducting arises from the formation of Cooper pairs, which act as carriers without resistance. The binding energy between two electrons of these pairs is weak; however, the correlation distance of two electrons (ξ) is long and can reach up to 10^{-4} cm, which is more than 104 times that of lattice size based on the calculation of second-order phase transition theory in a superconductor. Therefore, the superconducting correlation is a long-range interaction and can occur in space spanning many lattices. Further, these are possible for many Cooper pairs occupying the same space.

By introducing non-local electrodynamics into superconductivity, Pippard proposed the concept of superconducting coherence length ξ. According to the London equation, the penetration depth λ of a superconductor is constant and depends on material properties, as well as on temperature. Based on the Ginzburg–Landau theory and experimental results, correlations on the London penetration depth can be made.

$$\lambda(T) = \lambda(0)\left[1 - \left(\frac{T}{T_c}\right)^4\right]^{-1/2} \tag{1.8}$$

where $\lambda(0)$ refers to the penetrating depth of superconductors at 0 K. Table 1.1 lists the penetration depth for several superconducting materials at 0 K. Theories and experiments show that the superconducting coherence length relates to temperature and is approximately given as:

$$\xi(T) = \xi(0)\left(1 - \frac{T}{T_c}\right)^{-1/2} \tag{1.9}$$

where $\xi(0)$ denotes the coherence length of a superconductor at temperature 0 K. The magnitudes of coherence length of several superconductors are also presented in Table 1.1.

TABLE 1.1

Macroscopic and Microscopic Characteristic Parameters of Several Superconductors [5]

Superconductors	Crystal structure	Lattice constant Nm	T_c K	λ (0 K) nm	ξ (0 K) mm
NbTi;	A2		9.3	300	4
V_3Ga	A15		15	90	2~3
V_3Si	A15		16	60	3
Nb_3Sn	A15		18	65	3
Nb_3Al	A15		18.9		
Nb_3Ga	A15		20.3		
$Nb_3(Al_{.75}Ge_{.25})$	A15		20.5		
Nb_3Ge	A15		23	90	3
NbN	B1		16	200	5
V_2 (Hf, Zr)	C15		10.1		
$PbMo_6S_8$	Chevrel		15	200	2
MgB_2	Hexagon		39	140	5.2
$La_{1.85}Sr_{0.15}CuO_{4-\delta}$	I4/mmm	0.3799	40	80 (ab), 400 (c)	~4(ab), 0.7(c)
$YBa_2Cu_3O_{7-\delta}$ (YBCO, Y123)	Pmmm	0.3884	90	150 (ab), 900 (c)	~2(ab), 0.4(c)
$Bi_2Sr_2CaCu_2O_{8-\delta}$ (Bi-2212)	A2aa	0.542	90	300 (ab)	~3(ab), 0.4(c)
(Bi, Pb)$_2$Sr$_2$Ca$_2$Cu$_3$O$_{10+\delta}$ (Bi-2223)	Perovskite (Orthogonal)	0.54	110		
$Tl_2Ba_2CaCu_2O_{8+\delta}$ (Tl-2212)	I4/mmm	0.3856	110	215 (ab)	~2.2(ab), 0.5(c)
$Tl_2Ba_2Ca_2Cu_3O_{10-\delta}$ (Tl-2223)	I4/mmm	0.385	125	205 (ab), 480 (c)	1.3(ab)
$HgBa_2Ca_2Cu_3O_{8+\delta}$	Pmmm		133		1.42(ab)

Note: (ab) denotes the ab plane of crystal structure; (c) refers to the c-axis direction perpendicular to the ab plane. The symbol "~" means "close to".

1.5.2 Classifications

The experimental result shows evidence that some superconductors in magnetic fields do not allow the penetration of the magnetic field before they lose superconductivity, even though the magnetic field is more than its critical magnetic field. Further, other superconductors permit the penetration of the magnetic field into their partial regions. In such cases, the interior of a superconductor exhibits local interlacement with normal state and superconducting state simultaneously, which is called *mixed state*, even though their resistance remains zero. Hence, superconductors can be classified into two types [8]. According to the Ginzburg–Landau theory, superconductors can be classified into two categories based on the ratio of penetration depth λ to coherence length ξ, by defining the Ginzburg–Landau parameter κ as:

$$\kappa = \frac{\lambda(T)}{\xi(T)} \tag{1.10}$$

If $\kappa < 1/\sqrt{2}$, superconductors have positive interface energy and are called Type I superconductors. Conversely, if $\kappa > 1/\sqrt{2}$, superconductors have negative interface energy and are called Type II superconductors.

1.5.3 Type I Superconductor and Magnetization

Type I superconductors, also known as Pippard superconductors or soft superconductors, have only one critical magnetic field H_c. If the temperature is below the critical temperature ($T < T_c$) and external magnetic field ($H < H_c$), the superconductor is in Meissner effect and the magnetization is equal to the magnetic field strength H in the opposite sense ($M = -H$). When the external magnetic field strength (H) exceeds the critical magnetic field (H_c), the superconductor immediately turns to the normal conducting state. Figure 1.5 shows the magnetization of Type I superconductor, which is reversible, and there are only two states, i.e., the superconducting state and the normal state. The superconducting phase diagram is shown in Figure 1.6. Further, due to the transportation of current only within the thin layer (thickness of penetration depth) of the superconductor's surface and no current flowing through the entire body, a Type I superconductor has little practical value in power applications.

1.5.4 Type II Superconductor and Magnetization

A Type II superconductor has two critical magnetic fields, which are defined as lower critical magnetic field (B_{c1}) and upper critical magnetic field (B_{c2}), when subjected to a temperature below T_c. When the external magnetic field (H) satisfies $H < H_{c1}$, the superconductor is in Meissner state. When the external magnetic field is in the range $H_{c1} < H < H_{c2}$, the superconducting state

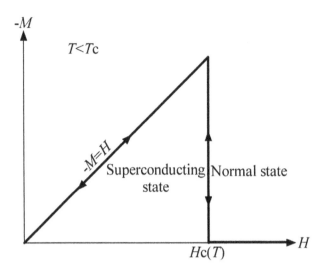

FIGURE 1.5
Magnetization (*M*) characteristic of a Type I superconductor under external magnetic field (*H*) at temperature (*T*) and less than critical temperature (*T$_c$*).

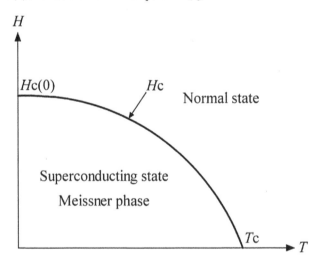

FIGURE 1.6
Magnetic phase diagram of Type I superconductor.

and normal state coexist and is known as *mixed state* as shown in Figure 1.7. Under such cases, flux lines can penetrate through the normal region inside the superconductor which is known as the flux vortex area. Figure 1.8 shows a Type II superconductor in the mixed state, in which the shaded areas denote normal region (N), through which the magnetic field can pass, while other parts are superconducting regions. Further, the magnetic phase diagram is represented in Figure 1.9.

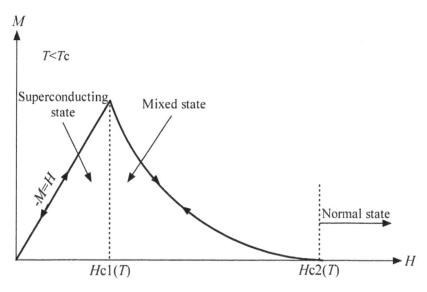

FIGURE 1.7
Magnetization characteristic of a Type II superconductor.

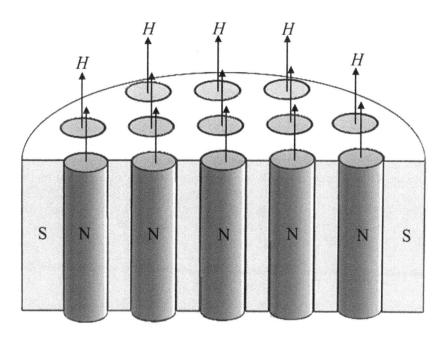

FIGURE 1.8
Type II superconductor in mixed state, which consists of superconducting state (S) and normal state (N).

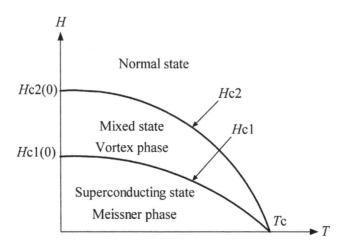

FIGURE 1.9
Magnetic phase diagram of Type II superconductor.

1.6 Superconducting Devices

High-temperature superconductors (HTS) have achieved a wide range of reliable technologies, among which the power system is one of the important applied field. In general cases, the superconducting apparatus of the power system is operated in a low magnetic field and liquid nitrogen (LN$_2$) temperatures. In recent years, several prototypes of many prototypes of HTS electrical power apparatus, such as transmission cable, fault current limiter (FCL), transformer, motors/generators, and magnets for energy storage devices, have been successfully demonstrated worldwide at both lab-scale and large-scale commercial projects. Such applications are presented in Figure 1.10. In particular, the field of electrical system has been most impacted by superconductors. Technologies, such as superconducting-cables, generators/motors, transformer, energy storage, and fault current limiters (shown in Figure 1.11), have shown the utility of superconductors. Hence, this section provides a brief overview of the superconducting devices for electrical power applications.

1.6.1 Superconducting Generator/Motors

A significant share of total power is consumed by electric motors in an industrialized country. Hence, the worldwide motive is to improve the efficiency of large motors. Likewise, power generators have also been in focus for improving their efficiency. Such motivations have been the driving force for the development of superconducting-generator/motors, ever

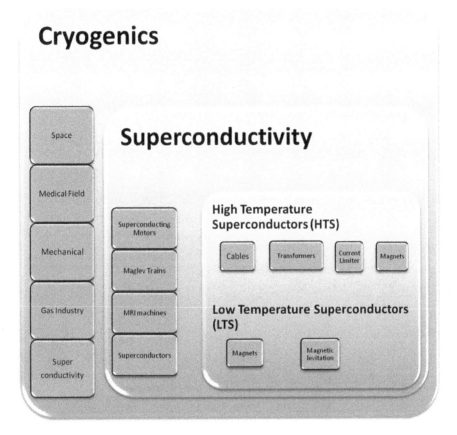

FIGURE 1.10
Application of cryogenics and superconductivity.

since the development of commercial superconductors. Such superconductor-based technologies have improved the efficiency, as well as reduced the overall weight of the system. Further detailed discussion is provided in Chapter 3.

1.6.2 Electric Energy Storage

The storage of electrical energy is a challenge, for which the electric grid has been designed accordingly. However, up to now, the main energy storage techniques have converted electrical energy into potential energy (pumped hydro-storage), high pressure (compressed gas storage), or chemical energy (batteries). A more direct means of electrical energy storage is in capacitor banks. Now, superconductors offer a new means of electric energy storage, with the loss-free circulation of electrical current in coils, thus generating magnetic energy, which is further discussed in Chapter 4.

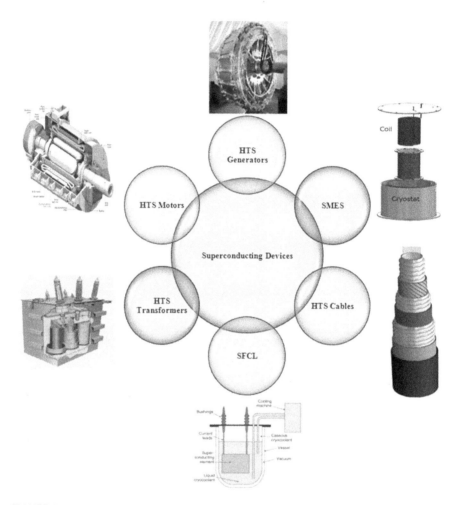

FIGURE 1.11
Primary devices for superconducting power grid: generators, SMES, fault current limiters, transformers, and motors.

1.6.3 Overhead Cables and Underground Cables

The most fundamental elements of the electric power grid are the long-distance links, which comprise of overhead cables and underground cables. Such lines or cables consist of long conductors arranged in an appropriate configuration to handle suitable voltage, current, and mechanical stress. Overhead lines, typically with aluminum wires, are suspended from wooden or metallic towers and are common throughout the world. Further, conventional underground cables consist of a central or "phase" conductor at their core, typically wound from multiple copper wires, which are surrounded by high voltage dielectric, such as cross-linked polyethylene. The whole assembly is protected with an outer jacket. However, the involved Joule loss in normal conductors provides a path for the superconductor to replace or retrofit such cables (refer to Chapter 5 for a more detailed discussion). Moreover, superconducting cables are laid underground, which are suitable for urban regions due to their lower right-of-way (RoW) requirements and lower weather-related damage, along with aesthetic concerns.

1.6.4 Circuit Limiters

Circuit limiters are an essential element of the electric grid, which is necessary to reduce the severity of fault current and isolate segments for repair or service in order to introduce new installations [9]. For such purposes, various switchgear equipment, circuit breakers, and electromagnetic relays have been developed to protect devices from electrical failures. Superconducting fault current limiters (SFCLs) have transformed the way electrical power systems function in terms of handling the fault current within milliseconds without interruption in the power demand, as will be discussed in Chapter 6.

1.6.5 Power Transformers

Another key element of the grid is the electric power transformer [9]. It is based on the principle of electromagnetic induction (Faraday's law). In a typical core-type transformer, two solenoid coils, namely, primary and secondary coils, separated by a dielectric insulator, are wound coaxially around a yoke which consists of a soft (easily magnetized) magnetic material. Due to the current in a link with resistance R generates Joule heating loss I^2R, long-distance transport of power with high current can be highly inefficient. In such a scenario, superconductors overcome this issue in a fundamental way, by bringing the wire resistance down to near zero, without any Joule losses. Further discussion of this is provided in Chapter 2.

1.7 Summary

This chapter has provided an overview of the fundamentals of superconductivity and related technologies developed for various engineering applications. The usage of superconductors has reformed the field of energy applications by creating reliable devices for globalized penetration in the energy market. However, fundamental challenges persist, which are being mitigated, and thus the upcoming chapters will discuss the superconducting technologies in the electrical system, by providing the reader with information on the current advancement, along with the issues pertaining to such systems, so that new ideas can be generated for developing new methodologies to be incorporated in superconducting systems.

References

1. H. Kamerlingh Onnes, "On the sudden change in the rate at which the resistance of mercury disappears," *Commun. from Phys. Lab. Univ. Leiden, 124c (Nov. 1911)*, pp. 818–821, 1912.
2. H. Rogalla and P. H. Kes, *100 Years of Superconductivity*. CRC Press, Boca Raton, Florida, 2012.
3. F. London, H. London, and F. A. Lindemann, "The electromagnetic equations of the supraconductor," *Proc. R. Soc. London. Ser. A - Math. Phys. Sci.*, vol. 149, no. 866, pp. 71–88, 1935.
4. J. Bardeen, L. N. Cooper, and J. R. Schrieffer, "Theory of superconductivity," *Phys. Rev.*, vol. 108, no. 5, pp. 1175–1204, 1957.
5. Y. Wang, *Fundamental Elements of Applied Superconductivity in Electrical Engineering*. Wiley, Singapore, 2013.
6. W. Meissner and R. Ochsenfeld, "Ein neuer Effekt bei Eintritt der Supraleitfähigkeit," *Naturwissenschaften*, vol. 21, pp. 787–788, 1933.
7. J. Ekin, *Experimental Techniques for Low-Temperature Measurements: Cryostat Design, Material Properties and Superconductor Critical-Current Testing*. Oxford: Oxford University Press, 2006.
8. C. P. Bean, "Magnetization of High-Field Superconductors," *Rev. Mod. Phys.*, vol. 36, no. 1, pp. 31–39, 1964.
9. H. W. Beaty and D. G. Fink, *Standard Handbook for Electrical Engineers, Sixteenth Edition*, 16th ed. New York: McGraw-Hill Education, 2013.

2

Cryogenic Cooling Strategies

Sudheer Thadela and Raja Sekhar Dondapati

CONTENTS

2.1 Introduction

The discovery of superconductivity in 1911 by Kamerlingh Onnes marked an era of advanced technologies, which had the potential of revolutionizing the field of power generation and transmission. He was measuring the electric resistance of metals at low temperatures and found that the resistance of mercury dropped to zero at the operating temperature of liquid helium (LHe) (4.2 K). Moreover, scientists Meissner and Ochsenfeld discovered that the magnetic flux completely disappears from the interiors of material when cooled to 4.2 K, which is the characteristic of perfect diamagnetism and is now termed "Meissner effect". The field of superconductivity has been attracting both theory- and experiment-oriented researchers to formulate a mathematical model and make reliable devices that can penetrate the commercial market at a global scale. Table 2.1 shows the superconducting technology which is of great value in the field of energy resources, power transmission, medical care, and large scientific instrumentations.

In 1986, a breakthrough was made by Karl Muller and Johannes Bednorz with the creation of brittle copper oxide ceramic compound, the so-called

TABLE 2.1

Superconducting Technologies for Power and Magnet Applications

Field	Application
Superconducting power technology	Generators/motors
	Current leads
	Superconducting magnetic energy storage (SMES)
	Transformer
	Superconducting fault current limiter (SFCL)
	Transmission power cable
Superconducting magnet technology	Magnet with high field strength
	Magnetic levitation (Maglev)

high-temperature superconductors, which exhibited superconductivity at 40 K. Since then, several high-temperature superconductors are continuously being fabricated, where the transition temperature has reached more than 90 K, which is higher than the operating temperature of liquid nitrogen (LN_2).

In the current scenario, nearly all the magnetic-grade devices for fusion applications, such as Large Hadron Collider (LHC), employ low-temperature superconductors (LTS) [1]. Further, magnetic resonance imaging (MRI) [2] that dominates the sector of medical diagnosis also employs LTS for generating a high magnetic field (0.2–0.3 Tesla). Whereas, high-temperature superconductors (HTS) facilitate fundamental changes in the power generation and transmission sector by replacing copper conductors with inherent high joule losses by superconductors, which eliminate resistive losses [3].

Despite current advancements, there are four major challenges HTS have to overcome to be widely used: cost, refrigeration/cooling, reliability, and acceptance. As this chapter deals with the cryogenic cooling strategies, the context will be focused on the refrigeration part for superconducting devices. However, readers can refer to other chapters of the book for more information on remaining issues. Cryogenics is a branch of physics that deals with temperatures below 120 K and is an essential component of a superconducting system; however, its involvement causes an increase in cost and complexity, which ultimately reduces reliability and safety. Hence, careful handling of cryogenic system is an essential criterion for the design of the superconducting system. This field is technically demanding and requires a multi-physics analysis by various engineering disciplines. Moreover, it is the combination of engineering and scientific disciplines, applied to realize low temperature, that makes process analysis, equipment design, and construction more challenging. It primarily involves the liquefaction of gases comprised in the atmosphere. Hence, the foremost challenge has been with regard to properties of pure chemical substances and their mixtures. The study of cryogenics involves many aspects, such as two-phase flows, supercritical fluids, and non-linear heat transfer characteristics. Moreover, the cryogenic system construction requires advanced knowledge of structural materials, assembly, joining techniques, heat-in-leak detection, along with the associated risk analysis under operation condition. This chapter aims to present the fundamental view of cryogenic cooling and associated equipment in relation to advanced engineering devices.

2.1.1 Cryogenic Heat Transfer Applications

The significance of heat transfer in cryogenic systems is illustrated by a variety of its applications. Some applications are discussed as follows:

1. Cryogenic fluid storage vessels (Dewars): For land-based service design of Dewars, the rate of heat transfer through the insulation must be as low as practical. In order to minimize the gaseous conduction, cryogenic insulations are generally evacuated. Moreover, multi-layer insulation (MLI) is used to minimize radiation heat

transfer, and low thermal conductivity spacers are used to minimize conduction through solids. In addition, the suspension system utilized for providing support to the inner vessel within the outer vessel must minimize the heat conducted through support. Hence, for vessel suspension purpose, high-strength stainless steel rods are often used due to its low thermal conductivity. For Dewars designed for space applications, the inner vessel may be shielded by both actively cooled radiation shields and vapor-cooled radiation shields. The radiation is intercepted by shields and transferred out of the system in order to minimize the evaporation rate of the cryogenic liquid. Moreover, glass epoxy and graphite epoxy are used to support the inner vessel within the outer vessel.

2. Superconducting magnetic energy storage systems: Conduction heat transfer along the current leads into the superconducting magnetic energy systems may be detrimental for the operation of magnet refrigeration system due to the associated thermal load. Therefore, such loads require special attention to design low heat transfer electrical feed-through [4].

3. Superconducting power transmission lines: The late 1960s marked the commercial development of superconducting electrical cables, which was made possible after the Nb-Ti and Nb_3Sn alloys became available in large-scale lengths [5]. Cables that are constructed from LTS were cooled to a temperature on the order of 4 K using LHe. However, the accompanied heat transfer caused extensive evaporation of LHe. Such a problem is solved by shielding helium circuits with LN_2-cooled thermal shields. Due to economic competitiveness compared with conventional ambient resistive power cables, cables based on LTS could be employed in applications where a large quantity of electrical power is to be delivered to a relatively small area. With the discovery of HTS, new interest emerged for developing superconducting energy transmission systems. HTS cables must be cooled to the cryogenic temperature, with the operating temperature of superconducting materials at ~80 K. Under such cases, the relatively inexpensive LN_2 can be used as a coolant. Such cables are generally covered with high thermal conductivity materials, such as copper, in order to reduce the severity of "fault current surges", where the superconducting current carrying medium momentarily changes to a normal conductor. Such transition causes the electric current to increase suddenly from 5 kA to as high as 100 kA for a short period of time.

4. Air separation systems: In an air separation system involving the production of LN_2, liquid oxygen (LOX), and liquid argon (LAr), high-performance heat exchangers are required. For the economical operation of these heat exchangers, the effectiveness must be approximately 95% or higher. Moreover, recuperative heat exchangers

allow the incoming warm air stream to be cooled by the outgoing cold gas stream, thereby reducing the need for external refrigeration. For reducing heat transfer from ambient surroundings, the cold heat exchangers must be further insulated. A simultaneous heat and mass transfer process occurs in the distillation column of the air separation system [6].

5. Micro-Electro-Mechanical System (MEMS) coolers: When cooled to cryogenic temperatures, the reliability and electronic noise characteristics of many miniature semiconductor electronic devices are improved. For achieving cooling over very small areas (spot cooling), micro-miniature Joule–Thompson refrigerators have been developed for such devices [7, 8]. These refrigerators involve heat exchangers which have flow channels of approximately 200 μm (0.008 inches) wide and 30 μm (0.001 inches) deep. A unique challenge posed by the design of such heat exchangers, due to modification of the governing equations, arises when the channel dimensions are on the order of mean free path of the flowing fluid [9].

6. Aerospace systems: The duration of aerospace missions may be as long as five years. Therefore, thermal protection for cooling instrumentation carrying any refrigerant liquid, such as LHe, must be provided. Such protection can be achieved by thermal shields, which are actively cooled by Stirling cycle refrigerators or thermoelectric coolers for maintaining an extremely low heat transfer rate to the coolant, thereby ensuring adequate mission duration [10, 11].

7. Cryosurgery systems: All practical surgical specialties, such as gynecological surgery, ophthalmological surgery, and neurosurgery, involve freezing a small region of the defective tissue for destroying the offending material [12, 13]. For such devices that involve the cooling process by either cryogens or small cryocoolers, thermal analysis of the tissue is of critical importance in order to predict the extent of the frozen lesion [14].

2.1.2 Development and Applications of Cryogenics for HTS System

In general cases, the demand for HTS systems is focused toward energy systems and health care. Moreover, national laboratories and industries across the world focused the attention toward the electrical system applications of HTS, recognizing the importance of cryogenics to enable such fields.

At present, the research base is focused on the following applications of HTS technology:

1. Industrial generator and motors (using Gifford McMahon (GM) single-stage cooler and pulse tube refrigerator (PTR)),

2. Transformer (GM coolers and LN_2 subcooling),

3. Cables (LN$_2$ subcooling and reverse Brayton cycle cooling),
4. Fault current limiter (GM single-stage coolers),
5. Magnetic storage devices,
6. Magnetic separation,
7. MRI,
8. Flywheel bearings (using GM single-stage and/or two-stage coolers and/or PTR).

In brief, the principal objectives are focused on:

1. Performance, cost, and reliability of the cryogenic refrigeration systems and
2. Development of the present state of cryogenic refrigeration systems, which are expected to match the needs of HTS power applications.

2.2 Cryogenic Cooling Methodology

The cooling of superconducting devices is an essential criterion for the chill-down of material in order to operate in the safety margin. Figure 2.1 shows the methodology for the cooling strategy of superconducting devices. The steps involved are as follows:

Step (I): Identification of superconducting devices – for the development and selection of cooling strategy, the initial step is to identify the device to be cooled.

Step (II): Selection of superconducting material – based on the targeted device, the superconducting material is to be identified. Such a judgment is based on the characteristic parameters of a superconductor, i.e., critical current (I_c), critical external magnetic field (H_c), and critical temperature (T_c) curve. For example, for magnet-oriented superconducting devices, the primary consideration is to have higher value of H_c and J_c. Hence, under such cases, T_c can be suitably reduced for increasing H_c and J_c, as the volume enclosed by the critical parameters is constant. Hence, based on H_c and J_c, the superconducting material needs to be identified, which can sustain the desired value for H_c and J_c. In a similar fashion, superconducting devices need to have higher J_c and T_c; hence, H_c can be suitably reduced.

Step (III): Selection of cryogenic fluid based on thermophysical properties – the basic understanding of cryogenic liquid is an essential requirement for superconducting devices. Table 2.2 shows the

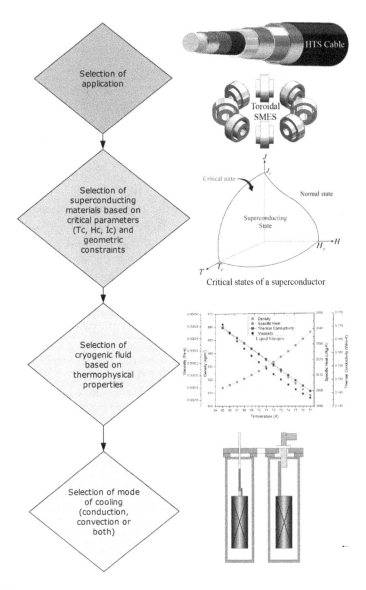

FIGURE 2.1
Cooling strategy for superconducting devices.

properties of common cryogenic fluids. The most commonly utilized liquid cryogens for superconducting devices are liquid nitrogen (LN_2) and liquid helium (LHe):

1. LN_2 is used to cool intermediate shield, in order to hinder heat leakages in the LHe medium, in other words, for pre-cooling purposes of LHe magnets to 80–100 K before achieving 4.2 K. Such

TABLE 2.2

Properties of Common Cryogenic Fluids

Cryogenic fluid property	He	H$_2$	Ne	N$_2$	Ar	O$_2$	CH$_4$
Critical temperature (K)	5.20	33.18	44.49	126.20	150.7	154.60	190.6
Critical pressure (atm)	2.25	12.98	26.44	33.51	47.99	49.77	45.39
Boiling point (K)	4.23	20.27	27.10	77.35	87.30	90.20	111.7
Melting point (K)	4.2	13.95	24.56	63.15	83.81	54.36	90.73

a process is economical from the point of view of LHe consumption compared to the magnet being directly cooled to 4.2 K from room temperature. In spite of development of superconductors having critical temperature above LN$_2$ operating temperature (~77 K at 2 bar), LHe still dominates the majority of high-field superconducting magnet applications.

2. LHe is used to operate superconducting magnets, fabricated using low-temperature superconductors (LTS).

Step (IV): Selection of mode of cooling – after the selection of cryogen, the next step is identifying the mode of cooling. Various methods are available, such as bath cooling or dry cooling (cryogen-free), and the selection from such methods is the capacity of the method to maintain the desired temperature in full loading or operational condition, i.e., considering all the heat sources such as alternating current (AC) losses and heat loads through external leakages.

There are primarily two methods for cooling a sample for cryogenic measurement:

1. The first methodology utilizes direct immersion of sample in cryogenic fluid. Such a method is useful when sample operation is at high current density. Moreover, the cooling is independent of shape (in wire, tape, or films, crystal or block). Further, direct cooling helps in achieving a high heat transfer coefficient between the cooling medium and the heat removal surface.

2. The second method is achieved using an intermediate medium, primarily by conduction cooling, where the sample is attached to a cold head of the refrigerator (cryocooler) as shown in Figure 2.2. The fundamental significance of such a methodology is that it minimizes the handling of cryogenic fluids. The immediate disadvantage after comparing with convection cooling is that this provides lower cooling power, which leads to limited transportation of current.

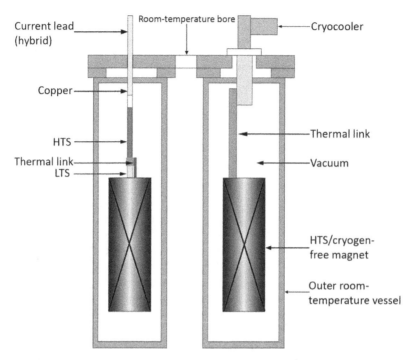

FIGURE 2.2
Schematic illustration of conduction cooled superconducting magnet system.

Moreover, for HTS applications, two forms of cryogens are available:

1. Using cryogens in solid form: There are three primary solid cryogens: solid nitrogen (SN_2), solid neon (SNe), and solid argon (SAr). Detail thermodynamic and mechanical analyses of the aforementioned solid cryogens are required before the selection of the desired HTS-based superconducting devices. SN_2 remains solid up to 64 K with good electrical insulation properties and is inexpensive, thus making it preferable. Therefore, for HTS magnets operating in the temperature range 20–60 K using BSCCO, YBCO, or MgB_2, SN_2 provides an enhanced thermal capacity, thus acting as a thermal reservoir. In general, solid cryogens possess good heating capacities and are suitable substitutes for impregnates using cryogens in liquid form.

2. Using liquid cryogens: Forced cooling using liquid cryogens provides a high heat transfer coefficient and is preferred in high-temperature superconducting devices for power generation and transmission application, such as superconducting motors and superconducting cables [15].

2.3 Cryogenic Cooling Strategies for Superconducting Devices

In the previous section, the various methods used for cooling of superconducting devices were discussed. Briefly, the following are the outcomes:

1. Bath-cooled: Mode of heat transfer is both conduction and convection [16].
2. Forced cooled: Mode of heat transfer is convection, such as in cable-
   in-conduit conductor (CICC) [17].
3. Cryocooled: Mode of heat transfer is conduction [18].

A block diagram of a typical cryogenic refrigerator is shown in Figure 2.3. A
heat load is present at the cold temperature (critical temperature T_c) at which
the HTS device is operated. This heat load could be from the environment
(typically radiation), solid/gas conduction, and fluid convection from adjacent
regions or structures, which are at a higher temperature. Moreover, the heat
load can also be generated in the superconducting components due to AC
losses. The heat from the load is transferred to the cold region of the refrigerator by solid conduction or fluid convection. The working fluid in the cold section absorbs this heat and flows to a heat exchanger, where the fluid is further
heated by exchange with a gas flow stream. After the process of heat exchange,
the fluid enters the compressor, where external work is done on the fluid in
order to obtain high pressure, thereby increasing the fluid temperature. After
this, the fluid flows to an ambient heat exchanger unit, where the heat absorbed
from the thermal load, and due to compression, is rejected to the ambient near
room temperature (T_a). Further, the fluid enters a heat exchanger, where it is
cooled by gas stream or a regenerator matrix. However, further reduction of
temperature can be achieved through Joule–Thompson expansion if the working fluid temperature is below the inversion temperature (~45 K for helium).

2.4 Heat Sources

The fundamental understating and calculation of heat sources is an essential procedure for the safe operation of superconducting devices. In order to
address the issues related to cryogenic or general cooling strategies, special
attention is required for the estimation of heat sources.

2.4.1 Conduction Heat Transfer at Low Temperature

Heat conduction mechanisms are a major concern while designing a cryostat
for testing or for commercial application. Hence, the efficient heat transfer

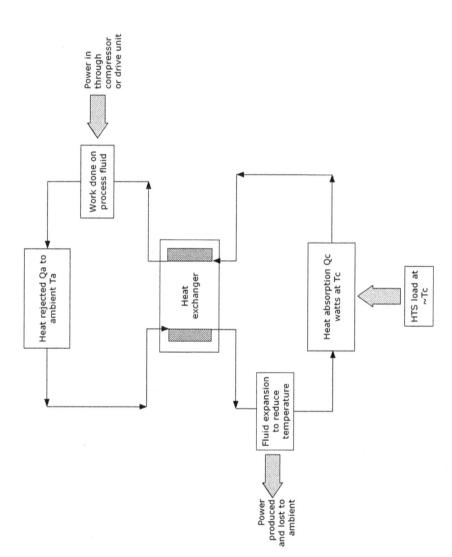

FIGURE 2.3
Block diagram of a typical cryogenic refrigerator.

from the sample or test environment across solid-solid link requires good thermal anchoring. Heat conduction is a phenomenon, which occurs through the interaction of neighboring atoms and molecules, resulting in the transfer of momentum. Therefore, thermal contact conduction between solid bodies in frictional contact is vital for cryogenic applications that are required to cool the system using cryocoolers. The steady-state conduction is achieved when the temperature difference driving the conduction is constant, which is obtained when the temperature in the heat-interacting conductor and sink does not change with time. However, in actual practice, the temperature within the conducting objects changes with time, resulting in transient heat conduction. In such a process, analysis is complex and is mostly carried out using empirical relations. Thus, limited measured data is available. Heat conduction through gases is utilized for shielding purposes. The heat transfer rate between the solid and gas (q_{conv}) is expressed by Newton's law of cooling (Equation (2.1)), where h is the convective heat transfer coefficient and A is the surface area.

$$q_{conv} = hA\Delta T \tag{2.1}$$

2.4.2 Radiation Heat Transfer

Due to the large temperature difference between the cryogenic medium and surrounding medium and components, a significant mode of heat transfer is experienced due to radiation. The radiation heat transfer is expressed by Stefan–Boltzmann law (Equation (2.2)), where σ is the Stefan–Boltzmann constant ($5.67 \times 10^{-8}\,W/m^2\text{-}K$) and ε is the emissivity of surface.

$$q_{rad} = \sigma \varepsilon A \left(T_{hot}^4 - T_{cold}^4 \right) \tag{2.2}$$

In order to minimize radiation heat leakages, an additional shield (referred to as radiation shield) could be place, which decreases the temperature gradient, thereby reducing radiation heat transfer into low temperature surfaces.

2.4.3 Joule Heating Due to Transport Current

The performance of a cooling system is affected by the transport current in the conductor, which results in resistive heating, also referred to as *joule heating* under AC conditions. Such heating significantly affects the performance of superconductors and needs to be compensated by the cryogenic system.

2.5 Cooling Sources

A cryostat is employed for achieving the desired cryogenic environment. A cryostat comprises of a double-walled evacuated vessel, which is cold within and stationary, similar to the Dewar flask, in order to maintain cryogenic temperatures, as shown in Figure 2.4. Such low temperatures are maintained using a refrigeration method or cryogenic fluids, depending upon the specification. Moreover, such vessels also act like pressure vessels during risky and hazardous circumstances. The typical size varies from cryocans utilized in laboratories for medical research to fusion reactor vessels for keeping the magnet cool at its operating temperature. The construction of a bath cryostat is similar in construction to a vessel filled with cryogens. A cold plate is in thermal contact with the cryogen, which is replenished as it boils away. Moreover, the boiling of cryogens, such as LHe, can be further used to cool the thermal shields placed outside the LHe bath. The boiling rate can be structurally controlled by installing several concentric layers of shield, which would gradually decrease the temperature. A closed-cycle cryostat consists of a chamber through which helium vapor is pumped and an external mechanical refrigerator that extracts helium exhaust, which is further cooled and recycled. Such a cryostat does not need refilling with helium; however, it demands large amounts of electric power. Within these cryostats, the magnet or the specimen is cooled by attaching a link, which is in thermal contact with it, to a metallic cold plate of a cryocooler. However, for laboratory experiments, continuous-flow cryostats have gained more attention, where LHe cools it from a storage Dewar flask. Continuous replenishment of LHe takes place by its steady flow. Such configurations do not require electric power; however, its drawback is the large quantity of LHe consumed during the operation. Hence, such facilities are encouraged to capture helium vented through the cryostat.

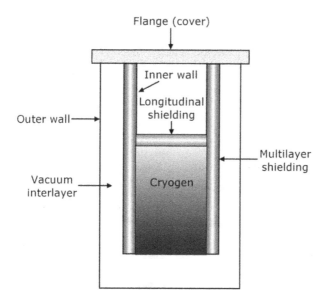

FIGURE 2.4
Schematic representation of a cryostat.

2.5.1 Cryogen

A cryogen is mainly utilized as a cooling source for cryogenic devices. The boiling temperature of cryogens is much lower than room temperature. The commonly used cryogens include liquid helium (LHe), liquid nitrogen (LN$_2$), liquid hydrogen (LH$_2$), liquid neon (LNe), liquid oxygen (LOX), etc. Table 2.3 shows the relevant thermodynamic properties of various cryogens. Due to the flammable and explosive nature of LH$_2$ and LO$_2$, rare and expensive features of LNe and LHe and LHe and LH$_2$ are generally selected as cryogens for superconducting power applications. Furthermore, LN$_2$ is rich in air, of which it occupies 78%, and is cheap and environment-friendly, making it an optimized cryogen in HTS power applications. He-3 and He-4 are the two isotopes of the helium family. The content of He-4 in air is less than 0.005%, while the content of He-3 is even less. Moreover, the boiling temperature of a cryogen must be lower than the critical temperature of the superconductor, if it is selected as a cooling source. Also, the boiling temperature of LHe (4.2 K) is lower than the critical temperature of LTS materials, MgB$_2$, and iron-based superconductors; hence, LN$_2$ and LHe can be used as cryogens for HTS and LTS materials, respectively. In most cases, the operating temperature of cryogens is kept lower than the corresponding boiling temperature of the cryogens due to the unaccountable heat leaks from surroundings.

2.6 Typical Guidelines for Cryogen-Free Magnet Systems

The cryogen-free magnet is cooled and maintained at its operating temperature by a closed-cycle cryocooler, which comprises of cold head, electrically

TABLE 2.3

Thermodynamic Properties of Several Cryogens at 1 atm and Critical Temperature [19]

Cryogenic fluid	Boiling point (K)	Density (kg/m³)	Specific heat (kJ/kg-K)	Thermal conductivity (W/m-K)	Viscosity (Pa-s) × 10⁻⁵
Helium-4	4.2304	124.73	5.2985	0.018656	0.31604
Hydrogen	20.277	70.797	9.6667	0.10341	1.3305
Neon	27.104	1207	1.8621	0.15500	11.610
Nitrogen	77.355	806.08	2.0415	0.14581	16.065
Argon	87.302	1395.4	1.1172	0.12565	27.071
Oxygen	90.188	1141.2	1.6994	0.15161	19.582
Methane	111.67	422.36	3.4811	0.18387	11.677

driven compressor, and connecting gas lines. The cryogen-free magnets are often designed with higher operating currents than equivalent wet magnets. The closed-cycle cooler may be of Joule–Thomson or pulse tube design, where both the arrangements consist of conduction cooling and does not require LHe or nitrogen. However, a cryogen-free magnet is generally bigger in size compared to magnets cooled by immersion in LHe. With wet magnets, the heat generated within coils, joint, and switches is accommodated with the high heat capacity available in LHe. However, in a cryogen-free magnet, the heat generated can only be transferred by conduction. In such cases, the only temperature gradient is in the direction of heat flow, which leads to lower heat transfer capacity in a cryogen-free arrangement. The operating temperature issues call for the following:

1. Systems designed for high operating temperature,
2. Coils with a higher current margin, due to generation temperature,
3. Higher permissible temperature at coils for better heat extraction (between coils and cooler),
4. Coping with the high temperature imposed by the rest of the system (excluding the magnet).

2.6.1 Design for Low Operating Temperature

The design of coils may be smaller; however, such coils result in reduced coil margins (T_G) which may result in quenching. Hence, the design of thermal extraction is very important in getting the desired cooling. During early AC loading conditions, a large amount of heat is generated. This heat must be removed at such a rate that the magnet recovers thermally before it reaches the high fields. The normalization of generated temperature takes place to assure that no section has an engineering J_c too high, in order to keep the total heat generation below the safety margin. Minimized heat generation and maximized cooling give rise to a low temperature field. Hence, suggestions for achieving the same are as follows:

1. Longitudinal heat generation gives direct route to end flanges with minimized eddy current heating,
2. Aluminum tape applied longitudinally gives conduction cooling and improves emissivity,
3. Material selection of high thermal conductivity,
4. Selection of wire, such as small filament diameter and short twist pitch,
5. High superconductor to non-superconductor ratio,
6. Heat generation rate lower than cooling capacity.

2.7 Elements of Construction

In this section, constructional elements required for the construction of superconducting devices are presented. Details regarding each component are discussed, along with the suitable recommendations for the instrumentation, such as thermometry.

2.7.1 Materials

The selection of material for cryostat construction is a challenge of its own. With the known material properties, there are limitations of what can be utilized for cryostat construction as well as heat transfer medium at cryogenic temperatures. Properties covering the requirement for thermal, mechanical, and electrical design are crucial for cryostat construction. Copper has excellent thermal conductivity at low temperatures, which makes it suitable for sample holders, for carrying heat to the sink. Moreover, most of the metals and non-metals become brittle at cryogenic temperature with the exception of face-centered-cubic (FCC) crystal structures, such as copper, brass, and aluminum. FCC materials generally show an increase in their yield strength at cryogenic temperatures. However, other non-metals do not embrittle, such as Kapton™ and Teflon™.

2.7.1.1 Thermal Properties

A crucial role is played by thermal conductivity in the selection of materials. Materials such as copper, brass, stainless steel, glass, epoxy composites, Kapton™, and Teflon™ of various grades are widely used for cryogenic applications. In general cases, thermal conductivity depends on the purity of the material. At cryogenic temperatures, higher purity metals possess higher thermal conductivities. In metals, purity is represented in terms of residual resistance ratio (RRR) of standard temperature and pressure (STP) to 4.2 K (RRR→R293 K/R4.2 K). At room temperature, the thermal conductivity in metals is mostly due to the vibration of lattice; however, temperature below 77 K is dominated by the effects of defects and impurities.

In dry systems, obtaining the best of thermal conduction and electrical isolation is a challenge (where the switches and joints sit in a dry environment and carry full transport current). The available options are to use sapphire and/or beryllium oxide as a medium to isolate electrically from the cooler stage/heat sink. Low thermal conductivity components in cryostats are generally made of austenitic stainless steel (AISI 304, 306, 321 of various compositions) for cryostat structures, which are occasionally used for supporting heavy structures, together with Cu-Ni (0.1) and Inconel 625 due to their mechanical strength at any cryogenic temperature and low thermal conductivity. For magnetic measurements, titanium is preferred over stainless steel and AISI 321. For support structure, non-metallic materials are used due to less conduction heat leakages. However, due to low thermal conductivity, the chill-down time is longer compared to metal/metal alloys.

1. Thermal contraction: This parameter is important with regard to construction and is required to be closely matched to avoid the stresses that would develop in the supporting structure. When cooled from room temperature to 77 K, the total thermal contraction in stainless steel, copper, brass, aluminum alloys, and epoxy resins varies substantially. Such a contraction is usually measured and defined in terms of $\partial l/l$ ($=L_{295\,K} - L_T/L_{295\,K}$) (unit/unit) or percentage contraction [20]. The qualitative criteria of thermal contraction is classified as high ($\partial l/l > 1\%$), medium ($\partial l/l > 0.3\%$), and low ($\partial l/l < 0.3\%$).

2. Heat capacity: This is classified in terms of specific heat, which is a measure of the amount of heat required per unit mass to raise the temperature by 1 K. In general, at cryogenic temperatures, the heat capacity decreases by an order of T^3. Hence, at cryogenic temperatures, some materials take longer time to cool down and reach a stable state.

2.7.1.2 Mechanical Properties

Mechanical properties are of primary concern for structural materials, and materials that embrittle on cooling are not utilized. Such considerations limit the use of most alloys, such as carbon steel and Fe-Ni steels. FCC structured materials, such as copper, aluminum, and brass, are used due to their excellent ductile properties at cryogenic temperatures. However, titanium hexagonal closed pack (HCP) and niobium for shielding body-centered cubic (BCC) can be used with careful design on thermal stress and strain. Moreover, yield strength along with magnetic susceptibility is also considered while designing the structural materials for room and cryogenic temperatures.

2.7.1.3 Magnetic Properties (Magnetic Susceptibility)

For selection of materials for magnetic environment, magnetic susceptibility is one of the factors that play an important role. In the presence of high magnetic fields, the forces induced could be significant, owing to the choice of materials. The magnetic susceptibility of stainless steel is reasonably high when exposed to thermal cycles and thus cannot be neglected during the design stage. Table 2.4 shows the materials used in a low-temperature environment.

TABLE 2.4

Preferred Materials in Low-Temperature Environment [21]

Material	Comments
SS 321	Not recommended for magnets
SS 304	Most commonly used for both coil support and cryostat
SS 304L	Commonly used (preferred for magnet/coil support and cryostat)
SS 316	Preferred for magnet vessel/cryostat
SS 316L	Preferred for magnet vessel
SS 316LN	Highly recommended for magnet vessel
Aluminum-1200(0)	Pure aluminum. Radiation shield and neutron beam applications
Brass-CW712R, CW507L	Most commonly used in magnet assembly joints, leads, structure
Copper-CW004A	Most commonly used
Copper-CW008A	Oxygen free high thermal conductivity copper (UHV application)
EP CC 301, 6F45, EP GC201	Standard sheet used for magnet, epoxy/cotton, epoxy/glass

2.7.2 Vacuum

The presence of vacuum is essential at all times for containing cryogens within a cryostat in order to reduce the heat load, either through conduction or convection, from room temperature. Hence, vacuum forms the primary element for achieving the superconducting state of materials, as shown in Figure 2.5. Moreover, the level of vacuum requirements varies depending upon the application, as shown in Table 2.5. The rate of desorption of gas trapped on the surface of materials is extremely low, due to which most of the vacuum spaces are evacuated before cooling and subsequently sealed.

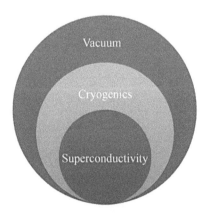

FIGURE 2.5
Primary aspect for construction of superconducting devices.

TABLE 2.5

Classification for Degree of Vacuum

Degree of vacuum	Pa
Rough vacuum	~99000–3300
Medium vacuum	~3300–1 × 10^{-1}
High vacuum	~10^{-1}–10^{-4}
Very high vacuum	~10^{-4}–10^{-7}
Ultra-high vacuum	~10^{-7}–10^{-10}
Extreme high vacuum	<10^{-10}

Moreover, most of the cold surfaces act as cryo-pumping surfaces, which lower the partial pressure of most gases in the system, except helium, hydrogen, and neon. In addition to the selection of a suitable vacuum pump, as shown in Figure 2.6, proper vacuum ducting is essential. Another significant factor in vacuum design is to limit the total gas influx rate to a consistent level while maintaining the ultimate pressure required of the vacuumed space. The sources of gas in vacuum are mentioned below:

1. The leakage of vacuum vessel generally occurs at joints (temporary and/or permanent). For high vacuum (HV) and ultra-high vacuum (UHV) systems, the selection of material is of prime importance. Moreover, the materials are generally heated to a temperature >400°C for removing any adsorbed gas within the vacuum chambers surface. Therefore, welding and hard soldering (brazing) are preferred in HV and UHV systems.

2. Virtual leak is normally due to trapped gases within the vacuum system. Such a condition is critical because this limits the ultimate pressure in the vacuum space and prolongs the evacuation time. Hence, it is always suggested to have a continuous weld on the vacuum side.

3. Degassing/out-gassing of materials is always of primary concern for any vacuum system. Most of the surfaces in the vacuum system contain adsorbed gases (surface phenomenon), such as moisture and gas molecules. Such issues are not a major concern for rough- to medium-vacuum systems; however, it significantly affects HV and UHV systems. Therefore, materials with rough surfaces, which provide substantial surface area, oxidized surfaces, insulated wires, etc., are difficult to use in HV and UHV systems. Hence, to avoid out-gassing over time, vacuum chambers are degassed by baking the vacuum system while maintaining it under vacuum conditions [22]. Though it is proven that vacuum chambers are needed to be heated in order to remove the adsorbed gas molecules, however, the question still remains: To what temperature and duration should heating take place? Such a process is dependent on the binding energies

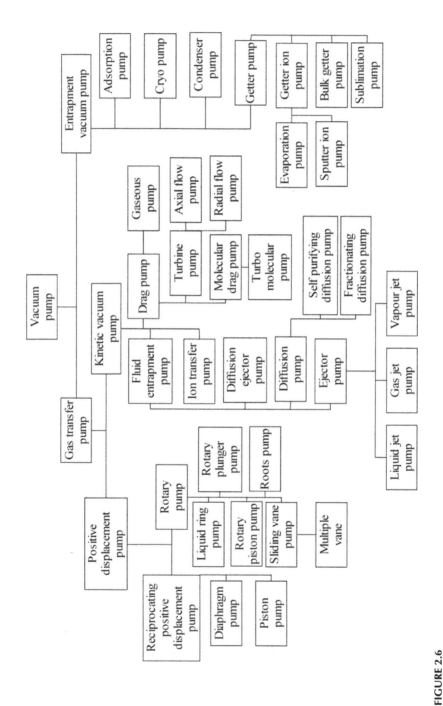

FIGURE 2.6
Classification of vacuum pump [23].

between the molecules and the material of the vessel/cryostat. It has been shown that preheating between 100–150°C and about 400–450°C for UHV systems provides desirable results in achieving the desired vacuum level.

4. Permeation of gases is a major concern for cryogenic vessels, which are usually evacuated at room temperature and sealed before cooling in order to reduce the pressure significantly. Hence, its mechanism is essential for the detailed calculation of the vessel/container thickness.

2.7.3 Cryogenic Insulations

The study of cryogenic insulation is of significant importance which needs special attention. Hence, a separate section has been devoted to provide a detailed discussion on this subject of interest. There are several types of insulation uses in a cryogenic environment, which include (1) expanded closed-cell foams, (2) fibrous materials and gas-filled powders, (3) aerogel insulation, (4) vacuum, (5) evacuated powders, (6) opacified powders, (7) microsphere insulation, and (8) MLIs. The characteristics of each type of insulation are discussed in the following sections.

2.7.3.1 Expanded Closed-Cell Foams

Expanded foam insulations have a cellular structure, which is produced by evolving gas during the formation of the foam. Foam used in cryogenic insulation includes polystyrene foam, polyurethane foam, and glass foam. Based on the type of gas used in the manufacture of foam, the density of the foam and the mean temperature decide the thermal conductivity of the foam. Moreover heat transfer through the foam is also influenced by convection and radiation within the cells and conduction through solid structure. The typical thermal conductivity values for foam are shown in Table 2.6. The following theoretical model had been developed [24] in order to predict the thermal conductivity of closed-cell foams:

$$k_t = 0.4(1-\varphi)k_s + k_g + 4F_e d\sigma T_m^3 \qquad (2.3)$$

where,

φ is the porosity of foam $\left(\varphi = \dfrac{\text{Volume of void}}{\text{Total volume}}\right)$,

k_s is the thermal conductivity of solid foam material,
k_g is the thermal conductivity of gas within the cavities,
d is the thickness of foam,
σ is the Stefan–Boltzmann constant (5.67×10^{-8} W/m²-K⁴),
T_m is the mean absolute temperature $\left(\frac{1}{2}(T_H + T_C)\right)$, and
F_e is the emissivity factor.

TABLE 2.6

Density and Thermal Conductivity of Selected
Foams and Gas-Filled Insulations [25]

	Density	Thermal conductivity
	(kg/m³)	(mW/m-K)
Foam insulation		
Polyurethane	11	33
Polystyrene	39	33
Polystyrene	46	26
Silica	160	55
Glass	140	35
Powders		
Perlite	210	44
Vermiculite	120	52
Aerogels		
Silica aerogel	80	19
Nanogel pack	116	15
Fibrous materials		
Fiberglass	110	25
Rock wool	160	35

$$\frac{1}{F_e} = \frac{d}{d_c}\left(\frac{1+r-t}{1-r+t}\right)+1 \tag{2.4}$$

where,
  r is the reflectivity of foam material,
  t is the transmissivity of foam material, and
  d_c is the size of foam cells.

The typical values of reflectivity and transmissivity of polyurethane foam
are 0.034 and 0.493, respectively. Moreover, cell sizes of the order of 0.25–0.50
mm can be easily achieved. The mean apparent density of the foam is related
to porosity and is given by Equation (2.5):

$$\rho_m = \rho_s(1-\varphi)+\rho_g\varphi \tag{2.5}$$

The first term in Equation (2.3) indicates the energy conducted through the
solid structure of the foam, the second term accounts for the energy conducted
through gas in the voids of the foam, and the last term represents the energy
radiated across the void spaces in the foam. Due to the small size of cavities in
the foam, convection is suppressed in the gas, which leads to only the gaseous
conduction mode being operative. Since the porosity of many types of foam is

near unity, the dominant mode of heat transfer is conduction through gas. In the past, refrigerant gases, such as R-11 and R-12, were used as foaming agents because of their low thermal conductivity and slow diffusion rate through the cell walls. However, these chlorofluorocarbons (CFCs) are responsible for the depletion of ozone layer, which consequently resulted in the discarding of CFCs in 1996, in conformity with the Montreal Protocol. Further, carbon dioxide has been used as a foaming agent because of its low vapor pressure at cryogenic temperatures. When the foam is cooled below the sublimation point of CO_2 (−109.4°F or −78.6°C), the gas freezes, which results in a slight production of vacuum in the cells, and the thermal conductivity of the foam is decreased. The diffusion of CO_2 is more through the cells compared to that of R-11 or R-12; thus, some of the CO_2 is replaced with air as the foam ages, leading to the increase of thermal conductivity by 40%. Moreover, if the foam is exposed to an atmosphere of hydrogen or helium for an extended period of time, the foam's thermal conductivity may increase by a factor of 3 or 4, because the thermal conductivity of H_2 or He is much higher than that of CO_2. Foam insulations are generally less expensive than other cryogenic insulations. When the cryogen cost is low, such as liquid natural gas (LNG) applications [26], the foam insulations present an alternative insulation, and therefore a higher heat transfer rate can be tolerated. The thermal drawback of rigid foam insulation is the large thermal contraction, which further gives rise to thermo-mechanical issues. The thermal expansion coefficient of polystyrene foam is $7.2 \times 10^{-5}°C^{-1}$ (temperature range between −30°C and +30°C), whereas the thermal expansion coefficient of carbon steel is only $1.15 \times 10^{-5}°C^{-1}$. For example, if the temperature difference is 100°C and a total length of 6.0 m, the difference in thermal contraction of foam relative to steel is 36 mm; thus the differential thermal contraction is not negligible. However, for preventing the diffusion of water vapor and air into the foam, the foam insulation requires a vapor barrier. Since the exposure to hydrogen and helium gas causes deterioration in thermal conductivity, the vapor barrier is also necessary to reduce the replacement of foaming agent with these gases.

2.7.3.2 Gas-Filled Powders and Fibrous Materials

Porous insulations include perlite (a silica powder), fiberglass, silica aerogel, vermiculite, and rock wool. The thermal conductivities of these materials are given in Table 2.6. The primary mechanism of thermal insulation in the gas-filled powders and fibrous materials is the suppression of gaseous convection due to the small size of voids within the material. Nusselt developed the following expression for the apparent thermal conductivity of a gas-filled powder insulation, which is given by Equation (2.6),

$$\frac{1}{k_t} = \frac{1-\varphi}{k_s} + \left[\frac{k_g}{\varphi} + \frac{4\sigma d_s T_m^3}{(1-\varphi)} \right]^{-1} \qquad (2.6)$$

where,
   d_s is the size of particles or fibers and
   φ is the porosity.

For gas-filled powders at cryogenic temperatures, the radiant contribution (the term involving T^3) is generally smaller than the gaseous contribution, and the apparent thermal conductivity approaches:

$$k_t = \frac{k_g}{1-(1-\varphi)\left(1-\dfrac{k_g}{k_s}\right)} \qquad (2.7)$$

Moreover, the thermal conductivity of solid material is often much larger than the thermal conductivity of gas within powder insulation. For example, the thermal conductivity of silica glass is 0.95 W/m-K and the thermal conductivity of nitrogen gas is approximately 0.017 W/m-K. Therefore, the ratio k_g/k_s is often negligible compared with unity and Equation (2.7) approaches:

$$k_t = \frac{k_g}{\varphi} \qquad (2.8)$$

The thermal conductivity of gas-filled powders and fibers approaches the thermal conductivity of gas in the insulation. An exception to this statement arises in the case of very fine powders, where the mean free path of the gas becomes larger than the mean spacing of powder particles. As indicated by Equation (2.9), the thermal conductivity of gas is proportional to the mean distance the molecules travel between collisions.

$$k_t = \frac{1}{3}\rho c_v \bar{v} \lambda \qquad (2.9)$$

The distance between molecular collisions will be reduced for very fine powders, which result in the decrease of effective gas conductivity, thereby decreasing the effective thermal conductivity of insulation. The thermal conductivity is generally decreased as the particle size is decreased, for fixed insulation density [27]. However, such decrease in thermal conductivity is also influenced by the overall contact resistance because more contact exists in series for small powder particles compared to larger particles.

2.7.4 Electric Wiring and Connections

The current topic is huge and challenging and thus this section is an attempt to describe the requirement that needs to be addressed while defining access into the operational environment (cold and/or vacuum). This section highlights the various types of wiring in cryogenic engineering:

- DC (direct current) and LF (<10 kHz) wiring. Such wiring is usually tightly twisted in pairs, in order to minimize the voltage pickup in a test environment. This method is more effective than co-axial cables for reducing magnetically induced noise.
- AC and HF (>10 kHz) wiring. The selection of wiring for a cryostat is primarily driven by the following: easy to work with (soldering), heat load, thermoelectric drift, and joule heating into the system, as shown in Table 2.7. The properties of wiring materials that are considered are physical properties, thermal conductivity, and RRR. The wiring provides a critical link from the cryogenic and vacuum environment to the outside world. Thus, the design of the electric feed is crucial from that of design consideration. Such feed-throughs are primarily used for instrumentation, diagnostics, and electrical power, for both cryogenic and non-cryogenic environments.

2.7.5 Thermometry

The measurement of temperature is essential in designing any cryostat system. For the evaluation of the cooling performance, the monitoring of conductor temperature, cryogenic fluid, and associated elements is an essential procedure. Due to the vastness of the current topic, we confine our discussion to the commonly used temperature sensors in superconducting systems

TABLE 2.7

Cryostat Wiring Recommendations

Application	Wire material
Voltage potential taps	Copper • For sensitive DC measurement • Preferred for continuous length
Heater	Constantan and phosphor bronze • Easy to solder and handle; however expensive
Thermometer leads	Phosphor bronze and copper Phosphor bronze is preferred over copper • Good thermal isolation • For long-length application, copper is preferred due to low resistivity
Low current leads	Copper and phosphor bronze • Copper preferred because of low resistivity
Heater leads	Copper • Low resistivity (less joule heating)

(primarily between 4 K and 300 K). In general cases, the following sensors are utilized:

1. Diodes: At a constant forward-biased current, the junction voltage of a semiconductor device increases with a decrease in temperature. The widely used sensors are based on Si and GaAlAs junctions.

2. Resistors: Resistance thermometers are available for both negative and positive temperature coefficients. For use in magnet field, negative temperature coefficient sensors based on carbon resistors, germanium, and carbon glass resistors are preferred over RuO. The use of positive temperature coefficient is limited to low-field applications only. Presently, Cernox™, a zirconium oxynitride, thermometers are used for a wide range of temperature measurement in the range of 4–300 K.

3. Thermocouples: Type E and Chromel-AuFe (0.07%) are the two primary thermocouples that have reasonable sensitivity in high magnetic fields and good signal level. For AC loss measurement, such sensors are generally preferred while employing thermal/temperature rise method. The challenge of using thermocouples is in its accuracy under the influence of magnetic fields. Moreover, along with sensor sensitivity, the wiring associated with the sensor contributes to signal-to-noise stability in an AC environment.

A wide temperature range of 4–1500 K is covered by type K thermocouple; however, it has half of the sensitivity compared to that of type E. Type T (Cu-Constantan) is preferred for cryogenic environments; however, its stability in cryogenic environments is a challenge due to high thermal conductivity from the extended copper arms of the sensor.

The challenge in cryogenic environment is the accuracy of measurement. The first issue is regarding the basic sensor accuracy and stability and the second is in the error seen due to wiring. In most cases, the primary reason for errors in measurement is due to the thermometer instrumentation and calibration. Hence, detailed analysis is required before the implementation of thermal connection between the sensor and sample/magnet. The following are good practices for any sensor wiring in a cryogenic environment:

1. High thermally conducting materials are to be used for anchoring the sample to sample/magnet.

2. Proper application of thermal grease should be attached to the holder.

3. For temporary sensor mountings, beryllium copper springs are to be used, along with thermal grease to the sensor holder.

4. The instrument leads serve as an intercept of the heat flow from elevated temperatures upon thermal anchoring. For sensitive measurements, a number of thermal anchor points are advised. Moreover, the final heat sink should be close to the sample temperature in order to avoid thermo-electromotive force across the leads.

5. For accommodating the strain effect between the sensor and the holder upon differential thermal contraction, the use of indium over Pb–Sn soft solder is advised.

Table 2.8 lists the sensor types and accuracy that are essential in deciding the measurement schematic and performance monitoring. For cryogenic system tests, the heat load to the system is usually measured based on pressure, temperature, and mass flow rate of input and output. In most cryogenic systems, the heat loads have been measured on large-scale systems, mostly covering the issues surrounding the accuracy of measurement. The application of first law of thermodynamics provides the energy balance for heat loads on

TABLE 2.8

Instrument List and Accuracy [28, 29]

Type	Range	Magnetic field application	Accuracy
Metallic resistance sensors (positive coefficient)			
Platinum	77–800	Yes, <0.1% error when used >70 K	±0.6 K at 70 K, ±0.2 K at 300 K (uncalibrated) 20 mK at 77 K, 35 mK at 300 K (calibrated)
Rh-Fe	0.5–900	No, <80 K	10 mK at 4.2 K, 25 mK at 100 K, 35 mK at 300 K (calibrated)
Semiconductor resistance sensors (negative coefficient)			
Cernox™	0.5–400	Yes, minimum error correction	5 mK at 4.2 K, 20 mK at 20 K, 50 mK at 100 K, 140 mK at 300 K
Carbon-glass	1–300	Yes, with correction	5 mK < 10 K, 20 mK at 20 K, 55 mK at 50 K (calibrated)
Diode voltage sensor			
Si diode	1.5–475	No, <70 K	20 mK between 1.5 K and 10 K, 55 mK > 10 K (calibrated)
GaAlAs diode	1.5–325	Yes, error less than Si diode but greater than Cernox™	15 mK < 20 K, 50 mK at 50 K, 110 mK at 300 K (calibrated)
Thermocouples			
Type E	3–1250	No	1.7 K from 75 K to 273 K
Chromel-AuFe (0.07%)	3–325	No	–

different components. The heat-load balance including the cryogenic system is given by Equation (2.10) and Equation (2.11).

$$Q_{total} = Q_{HTS-system} + Q_{cryogenics} = \dot{m}(h_1 - h_2) = \dot{m}c_p(T_1 - T_2) \qquad (2.10)$$

$$Q_{HTS-system} = Q_{conduction} + Q_{radiation} + Q_{joule-heating} \qquad (2.11)$$

2.7.6 Switch (in Case of Persistence)

A cryogen-free switch opened using a pulse to switch between the heater and the voltage to keep the switch open is conduction cooled. This can reduce the thermal load rather than having the current run through the switch heater continuously during the current ramp. Compared to a conventional wet switch, the cryogen-free switch cools from the inside diameter to the outside and acts as the heat sink. For HTS, persistence is still a challenge and development on this subject needs to address the joint technology.

2.7.7 Current Leads

The current leads are a crucial part of the overall system design and compete for space within a cryostat. A major connection is present between the magnet and the outside current carrying conductor, which introduces a significant amount of heat in the system. Since there is no gas cooling, such current leads are cooled by conduction only. Generally, current terminals at room temperature have resistive brass leads, which are connected to HTS leads at just above the level of cryocooler first stage (40 K). These brass leads are designed to give a best performance when carrying full magnet current, and the cross section and length can be adjusted to optimize its performance. Two heat loads must be considered: (1) joule heating due to current flowing through resistive brass and (2) heat conducted due to the temperature difference between the vessel and the cooler's first stage. Brass is used in preference to high-purity copper, due to less change in resistivity with change in operating temperature, which results in less change in thermal runaway. The other ends of the HTS are soldered to conventional copper-stabilized niobium titanium leads that link across to the magnet start and end joints (4 K). Usually the 40 K point is a solid joint so that heat conducted at room temperature and ohmic heating from the current being carried through the brass can be efficiently intercepted. Such a situation is complex, since the leads must be electrically isolated, along with good thermal contact with the cryocooler at the 40 K and 4 K points; hence, direct-bonded copper interfaces are used. The conducted heat is limited by using Au-Ag substrate HTS material mounted on brass shunt protection.

2.7.8 Protection

The design should be made assuming that the magnet and protection work together adiabatically during a quench. Moreover, the protection circuits should be designed conservatively so that they do not overheat. Further, oversize of wire would aid in cooling of the components by providing conduction paths.

2.7.9 Design and Manufacture of Joints

Joints present between wire grades and coils are likely centers of heat generation. In the LTS application, long-lap soldered tin joints and etched or spot welded Nb–Ti joints can minimize heat generation; however, for the HTS joint, technology is not yet advanced. Since HTS magnets are not persistent magnets, they can run on power supply and can be held across the power supply during operation. To conduct heat away from joints, use heavy short lead runs and thermal braids from start and end joints to the thermal bus (electrically isolated). If the magnet has a persistent mode switch, thermal interception is particularly important. Moreover, the use of joint cups, with large thermal mass for the start and end joints, may prove useful. Such joints will form a thermal buffer against heating from energized current leads and heat conducted from an open switch.

2.7.10 Thermal Shielding and Anchoring

A radiation shield is a part of thermal shield of the cryostat, which is fixed to the first stage of the cryocooler and enables it to run at a temperature of ~40 K. In large-scale magnets, a shield is normally applied around a magnet, joint, protection circuit, and current leads, which is anchored to the second stage temperature. Such a shield protects the second stage components from radiation, along with the electrically insulating material (Kapton™), in order to stop short-circuiting coil terminations, current lead, and protection circuits to each other or ground.

2.8 Cooling Strategies for Superconducting Devices in Electric Power Applications

In this section, the cooling strategies employed in superconducting devices are discussed. Based on the context of the chapter, we have focused on devises utilized for electric power generation and transmission.

2.8.1 Motors/Generators

The requirement for a cryogenic system in any superconducting motor or generator is of critical importance, which is also a major reason for non-commercialization of these machines. However, the situation has been eased with the development of HTS, which now can operate at 20–40 K, with the previous operating temperature of liquid helium (4.2 K). The primary issue is regarding the cooling of superconducting and armature coils below the T_c. Such cooling is achieved by circulating cryogenic fluid from a refrigerator to the coils and back in a closed loop. Though conduction cooling (cryogen-free) has been demonstrated for magnetic resonance imaging (MRI) machines [30], such cooling method is very difficult for rotating superconducting devices. For LTS machines, only the field winding is cooled, since cooling the superconducting armature would be expensive at 4 K. The circulation of LHe results in vaporization during the cooling of LTS field coil. Such a process leads to the development of the two-phase phenomenon due to its large temperature difference.

Figure 2.7 shows the configuration of a superconducting motor. Whenever the superconducting coil, either LTS or HTS, is placed on the rotor, the challenge of taking the cryogenic fluid on and off the rotor arises. Such issues are solved by providing a rotating cryogenic fluid coupling between the rotor and the stationary refrigeration system. Such a coupling must be operated at a cryogenic environment, along with thermal insulation. Operating at such conditions makes construction difficult; however, it has been achieved [31]. Moreover, it is possible to place the cold head of the cryocooler onto the rotor [32]; however, the electricity for power will still have to be conducted on and off the rotor at an ambient temperature. For HTS machines, more options are available, such as the operating temperature can extend up to ~40 K, with the

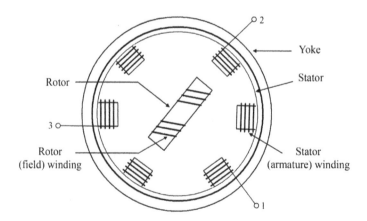

FIGURE 2.7
Configuration of a typical synchronous machine employing superconducting field winding on the rotor.

deployment of HTS wires (MgB_2). Though such HTS wires have transition temperature above the point of LN_2 at high pressures, such cryogens are still not usable due to the reduction of H_c with increasing T_c. Such circumstances open up the possibility of employing hydrogen (boiling point ~20 K at 1 atm) and neon (boiling point ~27 K at 1 atm) as additional cryogens. However, due to flammability and explosive characteristics, the selection of the former cryogen is a major concern. Nevertheless, the inert nature of neon has led to successful utilization in various prototypes [33]. Furthermore, mixed refrigerants have also been considered to cover intermediate temperatures [34]. In addition to such cooling techniques, gaseous helium may be circulated from the cryo-refrigerator to the HTS coils for maintaining temperature above 4 K, which eliminates the formation of two-phase flow. During LHe cooling, two-stage cooling is generally employed with Gifford–McMahon (GM) coolers in order to reach 4 K; however, for HTS applications, single-stage cooling is possible. Moreover, if cooling of HTS armature is required, cooling requirements will substantially increase and refrigerators such as reverse Brayton may be desirable [35]. With respect to cooling, another operating issue is regarding the chill-down duration. When the cooling is done from an ambient temperature, the refrigeration power incurred can take a couple of days to cool the devise to operating temperature. However, if the chill-down time is to be decreased, more refrigeration power is required; thus a trade-off exists between chill-down time and refrigeration power. Such issues can be addressed using a couple of approaches. The first one is to keep the machine cold by running the refrigerator continuously at non-operational hours of the machine. The second approach is the implementation of auxiliary tubes, carefully built and retrofitted during the development of superconducting motors. Such tubes can be fed with LN_2, which is relatively inexpensive compared to LHe and can cool the machine to an operating temperature of 77 K. Such an approach will minimize the cooling capacity for LHe-based cryo-refrigerator. In order to further minimize the cryo-refrigeration power, the system almost certainly requires vacuum insulation, which would eliminate the conduction and convection heat leakages from ambient. However, such insulation would further be a source of additional cost. Occasionally, foams have also been considered for insulation purposes for HTS operation, which are however less effective than vacuum. For LTS-based motors, emissivity reduction is required for reducing the heat leakage due to radiation. Such reduction involves the coating of low emissivity or blanketing the commercial MLI.

2.8.2 Superconducting Magnet Energy

HTS magnet-based applications are currently cooled by cryocoolers, where the cold head region is attached to one end of the magnet assembly, as shown in Figure 2.8. In general cases, the first stage of the cooler is thermally attached to the radiation shield and the second stage to the magnet assembly.

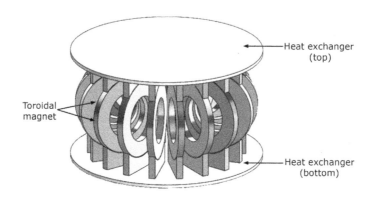

FIGURE 2.8
Toroidal-type SMES [36].

A major concern is to maintain a constant/minimum temperature gradient between the coldest region and the warmest point (assuming no additional heat load). Suppose the mass of the magnet be m, having heat capacity c, density ρ, volume (V), cross-sectional area (A_s), and Q'' represents the heat flux imposed on the conductor, then the total time (τ_{tot}) required to reach the desired operating temperature of superconductor can be calculated as follows:

$$\left(Q''_{leak}\right)A_s - \left(Q''_{AC}\right)A_s = \rho_S V_S c \frac{dT}{d\tau} \tag{2.12}$$

$$d\tau = \frac{\rho_S V_S c}{\left(Q''_{leak} - Q''_{AC}\right)A_s} dT \tag{2.13}$$

$$Q_{net}'' = Q''_{leak} - Q''_{AC} \tag{2.14}$$

$$\int_{0}^{\tau_{tot}} d\tau = \int_{T_i}^{T_f} \frac{\rho_S V_S c}{Q_{net}'' A_s} dT \tag{2.15}$$

$$\tau_{tot} = \frac{\rho_S V_S c}{Q_{net}'' A_s}\left(T_f - T_i\right) \tag{2.16}$$

2.8.3 HTS Cables

The thermal management of high-temperature superconducting cables (HTSC) is an important issue pertaining to its design and operation for eventually gaining acceptance in the electric grid. The operating range of an

HTSC typically ranges from 65 K to 80 K, which is determined based on the cryogen selected. LN_2 is the most preferable cryogenic fluid for HTSC operations, due to its inflammability, high heat carrying capacity, high dielectric strength, and abundance in nature. The minimum operating temperature is a balance between superconductors, which increases as the temperature is lowered, and which, increases the cost of refrigeration. With the usage of LN_2 as a coolant, the lowest temperature achievable for HTSC is 65 K (freezing point).

The choice of coolant used impacts the design of HTSC. These cables are cooled using LN_2 because in the desired temperature range of HTSC, LN_2 exits in liquid form. Figure 2.9 shows the temperature-entropy (T-S) diagram for LN_2, which shows that LN_2 will remain in subcooled region (liquid form) for a temperature below 87 K at 2.7 bars. However, coolants such as gaseous helium (GHe) and gaseous hydrogen (GH) have also been proposed [37]. Moreover, GHe has been proposed as a coolant for degaussing HTSC [38]. Further, the prospective use of gaseous coolants is in HTSC installation, where large changes in elevation occur due to the development of large pressure head. Such elevation head will significantly contribute to the total pressure, thereby increasing the pumping power for the system. Table 2.9 shows the comparison between the properties of LN_2 with gaseous coolants. The, usage of gaseous coolants will significantly reduce the circulation

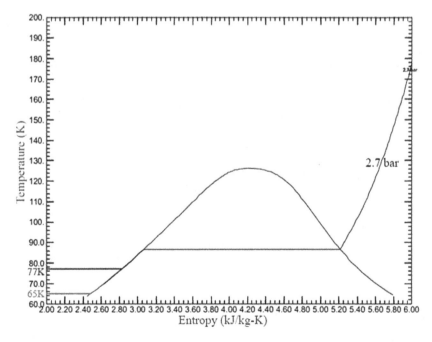

FIGURE 2.9
T-S diagram for LN_2 at an operating pressure of 2.7 bars.

TABLE 2.9

Cryogenic Fluid Properties at the Supply Condition of 0.3 MPa and Critical
Temperature K

Cryogenic fluid	Critical temperature K	Density (kg/m³)	Specific heat (kJ/kg-K)	Thermal conductivity (W/m-K)	Viscosity (Pa-s)
Liquid nitrogen	87.907	755.71	2.1176	0.12406	0.00010978
Helium-4	5	117.53	6.8544	0.020266	0.0000031139
Hydrogen	24.579	65.127	13.160	0.10418	0.0000096483

pump work due to the reduced density compared to liquid coolants. The cost
of the refrigeration system is based on the required cooling capacity. Hence,
an accurate estimation of heat load of HTSC and its accompanying terminations under full operational load is required. Therefore, minimizing the heat
loads will reduce the cooling capacity and, thus, the cost of cryo-refrigeration. Minimization of the thermal load is achieved by proper selection of
HTSC configuration and adequate thermal isolation from the surroundings,
along with the incorporation of vacuum-jacketed cryostat. Figure 2.10 shows
the various cooling schemes used for HTSC.

The HTSC is electrically connected at the ends to connections, which also
serve as a path from ambient temperature electrical connections to the HTS
cable. Moreover, under direct current (DC) load, no heat generation occurs in
the superconductors (if current transmitted is below I_c). The dominant heat
load stems from conduction through terminations and heat-in-leak from the
surrounding through the cryostat. However, under AC load, AC losses act as
a source of heat generation in the conductors and contribute to the total heat
generation.

Figure 2.11 shows the schematic of a single-phase cold-dielectric HTS cable.
The refrigeration system for HTSC can be divided into three categories: (1)
open cycle, (2) closed cycle, and (3) hybrid system. In the open cycle, the refrigerant is vaporized and then vented as shown in Figure 2.12. The open cycle is
the least expensive configuration in terms of initial cost; however, it requires
continuous replenishment of refrigerant. The second category is a mechanical
refrigerator, typically based on Brayton cycle, Claude cycle, or Stirling cycle
and a pulse tube system in future [39], which is illustrated in Figure 2.13. The
third category is the hybrid system, which uses a mechanical refrigerator as
the primary source for refrigeration and an open-cycle refrigerator as a back-
up, when the primary system is under maintenance or repair.

2.8.4 Superconducting Fault Current Limiter

A superconducting fault current limiter (SFCL) during normal operating
conditions, i.e., transports current lesser than the I_c due to the zero resistance.

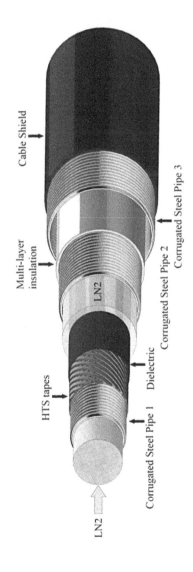

FIGURE 2.10
Illustration of various cooling schemes used for HTSC.

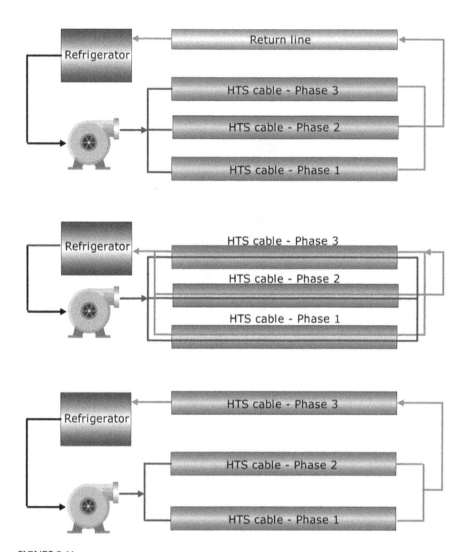

FIGURE 2.11
Schematic view of single-phase cold-dielectric HTSC (not to scale).

However, during a fault, the superconducting element rapidly changes and reverts to normal conducting state [40]. Under such conditions, the overcurrent flows through the shunt in connection, thereby protecting the end devices.

A genetic configuration of the SFCL cooling arrangement is shown in Figure 2.14, which includes an HTS element in a vacuum-insulated vessel with coolant, a cooling system (cryocooler), and a pair of current leads. The HTS element is connected in series with the line to be protected from the fault current. During a fault, the I_c is surpassed and the resistance of

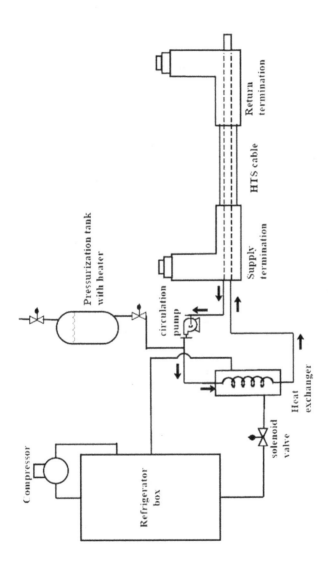

FIGURE 2.12
Vacuum-pumped subcooled system.

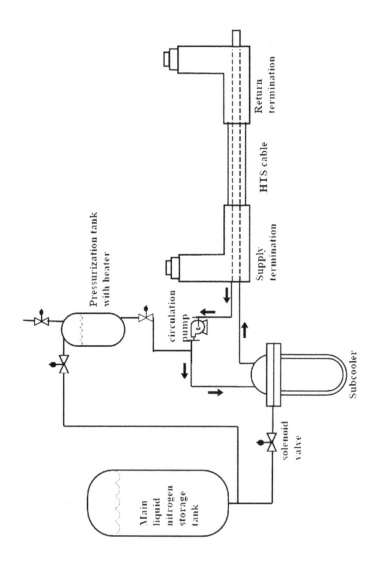

FIGURE 2.13
Integration of mechanical refrigeration system in the cooling of HTSC.

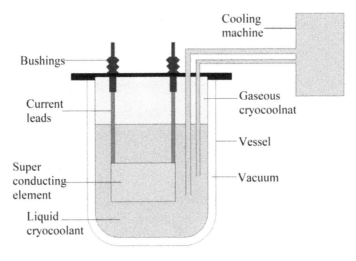

Bushings

Cooling machine

Current leads

Gaseous cryocoolnat

Vessel

Super conducting element

Vacuum

Liquid cryocoolant

FIGURE 2.14
Cooling arrangement of SFCL.

the superconducting element rapidly increases. Under such circumstances, the superconducting element acts as a heat generation source. Conduction and convective heat transfers occur (i.e., bath cooling) with the adjacent stationary cryogen, which prevents the quenching of the superconducting element. However, nucleate boiling of cryogen can occur, which can further be eliminated by the pressure of liquid cryogen or by cryogen present in gaseous form.

2.8.5 HTS Transformer

A practical cryogenic system for superconducting transformer must be quite, reliable, and efficient in operation. The required cooling power for such a transformer will be of several kilowatts of order at 70 K. Although conduction cooling had been attempted in the past [41], the recent developments are focused on the circulation of LN_2. The usage of other cryogens, such as helium, hydrogen, neon, and oxygen, is ruled out due to considerations of cost, safety, and temperature range. A circulated LN_2 provides consistent temperature distribution, along with rapid recovery from fault and the benefit of liquid dielectric. Various conceptual designs, such as configurations involving pumping mechanism [42] or circulation using thermosiphon [43], as well as bath cooling [44, 45] (Figure 2.15), have been demonstrated. Moreover, LN_2 offers dielectric performance compared to that of oil in a conventional transformer. However, at atmospheric pressure, the operating temperature margin of LN_2 is narrow (63–77 K). Hence, to prevent the formation of gaseous phase, thereby preserving the dielectric properties [46], it is desirable to maintain liquid nitrogen in a subcooled state in high-voltage

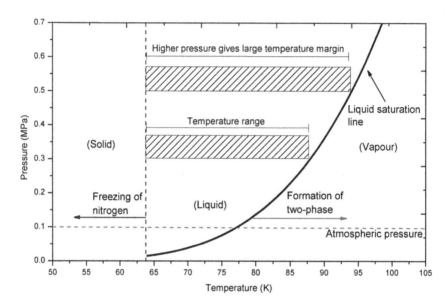

FIGURE 2.15
Liquid cooling system by natural convection for a superconducting transformer.

regions. Further, the cryogen must be pressurized to further increase the boiling point. Figure 2.16 shows the phase diagram of LN_2 showing the operating region suitable for a superconducting transformer. Stirling and GM cryocoolers are both proven technologies, which are capable of offering necessary cooling for a superconducting transformer. However, for larger rating transformer, multiple cryocooler units will be required [47]. A key disadvantage associated with Stirling and GM cryocoolers is that they require periodic and extensive maintenance. In such a scenario, turbo Brayton systems have been developed, which can deliver high efficiency and exhibit reduced maintenance, with a cooling capacity up to 50.8 kW [48]. Moreover, the cooling load will fluctuate as the transformer electrical loading wires varies; therefore it is necessary to vary the cooling capacity accordingly.

2.9 Issues Pertaining to Cryogenic Heat Transfer

1. Effects of variable material properties: At a cryogenic temperature range, the transport properties of materials generally vary significantly. For example, the specific heat of solids at low temperature varies as the third power of absolute temperature, whereas, the specific heats of metals nearly at room temperature may vary by less

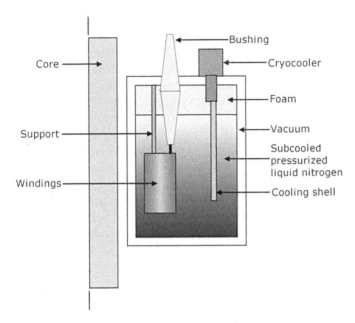

FIGURE 2.16
Phase diagram of nitrogen depicting the operating region for superconducting transformers.

than 5% for a temperature change of 100°F. Hence, constant property analysis may be valid for ambient temperature applications; however, such assumptions lead to inaccurate results when applied to cryogenic heat transfer applications.

2. Thermal insulation: All cryogenics liquids have a relatively small heat of vaporization. For example, the heat of vaporization associated with LN_2 is 199.3 kJ/kg at a pressure of 1 atm, whereas the heat of vaporization for water at 1 atm is 2257 kJ/kg. Considering safety instruction, liquefaction costs, and low heat of vaporization, special high-performance insulations are required in order to reduce the evaporation rate of cryogenic fluids in storage vessels. The thermal conductivity of multi-layer insulations (MLIs) utilized for cryogenic applications is approximately 1000× smaller than the thermal conductivity of fiberglass insulation commonly used in thermal insulations in residences.

3. Near-critical-point convection: For many cryogenic fluids (e.g., hydrogen and helium), the thermodynamic critical pressure is much lower than the critical pressure associated with conventional fluids (e.g., water). Hence, convective heat transfer at near-critical and supercritical conditions is more frequent in cryogenic systems.

4. Thermal radiation problems: According to Wien's law, the wavelength at which peak radiant intensity occurs for blackbody radiation

is inversely proportional to the absolute temperature. For example, at 1 K, the peak occurs at a wavelength of ~2.9 mm. Most metallic shields which are employed for reducing radiation heat transfer into cryogenic systems will have a thickness which is comparable to or less than this value. Hence, the treatment of radiation problems at cryogenic temperatures can be significantly different compared to materials at room temperature, where the wavelength peak occurs at ~0.009 mm.

5. Heat exchanger design: The liquefaction systems and cryocoolers are required to operate at high effectiveness levels, corresponding to only a few degrees of approach temperatures. Larger approach temperatures may be tolerated in more conventional applications, such as gas turbine heat exchangers or air conditioning heat exchangers. Hence, proper design of heat exchanger for cryogenic application is crucial for operation.

2.10 Challenges and Requirements

A major concern which limits the commercialization of cryogenic systems is the enhancement and expansion of the proven technology. Hence, the standardization of the cryogenic system is an essential requirement. Such a process will help in reducing the cost base and improve system reliability, rather than employing individual design to support each HTS system requirement. Another concern is the reduction of heat loads, primarily the static heat load into the vessel/cryostat containing superconducting components and dynamic load arising due to AC losses of the superconducting components under AC loading.

The current research effort is to support the commercial applications by HTS, thereby increasing the existing refrigeration technology above 30% Carnot efficiency. Such a target will focus research and development upon the improvement of compressor performance, significantly reducing static and dynamic heat loads on a case-to-case basis. Such coolers are evaluated in terms of coefficient of performance (CoP). CoP of a cryocooler is normally expressed as the ratio of refrigeration work output to the work of the compressor. The theoretical CoP of a GM cryocooler is given by Equation (2.17), which is based on the ratio of P-V work to the isothermal compression work,

$$\mathrm{COP_{GM}} = \frac{W_{P-V}}{W_{Iso}} = \frac{(P_{in} - P_{out})V}{mRT\ln\left(P_{in}\big/P_{out}\right)} \tag{2.17}$$

TABLE 2.10

Comparison of a Copper Magnet with Superconducting Magnet

Type of magnet system	Power (kW)	Energy required (continuous operation) (kW-hr)
Copper	2000	17520000
Superconductor	20	175200

where, $W_{P\text{-}V}$ is the *P-V* work per cycle, W_{Iso} is the isothermal compressor work (per cycle), P_{in} is the intake pressure, P_{out} is the exhaust pressure, V is the expansion volume, m is the mass flow rate (per cycle), R is the gas constant, and T is the compression temperature. Reducing the compression ratio of a real GM cryocooler improves CoP [49]. The running cost of a superconducting magnet is about 100 times less than that of equivalent copper magnet (low-field MRI system), as shown in Table 2.10.

2.11 Summary

The existing cooling schemes and technologies are proven; however, achieving cooling capacity for superconducting devices integrated in large-scale power applications is still a challenge. The foremost hindrance is to lower the refrigeration cost, thereby enabling superconductors to operate at a higher temperature. Moreover, for transmission purposes, AC losses are incurred which lowers the cryogenic efficiency. Hence, a gap is present to fully integrate or retrofit the present technologies with superconducting devices.

References

1. P. Lebrun, "Cryogenics for the Large Hadron Collider," *IEEE Trans. Appl. Supercond.*, vol. 10, no. 1, pp. 1500–1506, 2000.
2. T. Nakamura *et al.*, "Development of a superconducting bulk magnet for NMR and MRI," *J. Magn. Reson.*, vol. 259, pp. 68–75, 2015.
3. R. Wesche, A. Anghel, B. Jakob, G. Pasztor, R. Schindler, and G. Vécsey, "Design of superconducting power cables," *Cryogenics (Guildf).*, vol. 39, no. 9, pp. 767–775, 1999.
4. R. C. Niemann, J. D. Gonczy, and T. H. Nicol, *Low-Thermal-Resistance, High-Electrical-Isolation Heat Intercept Connection.* United States: Plenum Press, Springer, Boston, MA, 1994.

5. H. Rogalla and P. H. Kes, *100 Years of Superconductivity*. CRC Press, Boca Raton, Florida, 2012.
6. M. Grenier and P. Petit, "Cryogenic air separation: The last twenty years," in *Advances in Cryogenic Engineering*, Volume 31, R. W. Fast, Ed. Boston, MA: Springer US, 1986, pp. 1063–1070.
7. W. A. Litttle, "Microminiature refrigerators for Joule-Thomson cooling of electronic chips and devices," in *Advances in Cryogenic Engineering: Part A & B*, R. W. Fast, Ed. Boston, MA: Springer US, 1990, pp. 1325–1333.
8. B.-Z. Maytal and J. M. Pfotenhauer, *Miniature Joule-Thomson Cryocooling*. New York: Springer, 2013.
9. D. Y. Tzou, *Macro to Microscale Heat Transfer: The Lagging Behavior*. PA: Taylor & Francis, London, United Kingdom, 1996.
10. J. H. Lee, D. A. Payne, and R. D. Averill, "Design of the Cryogenic System for the Safire Instrument," in *Advances in Cryogenic Engineering*, R. W. Fast, Ed. Boston, MA: Springer US, 1991, pp. 1211–1219.
11. S. J. Nieczkoski, R. A. Hopkins, and S. R. Breon, "5-Year lifetime hybrid superfluid helium dewar for the AXAF X-ray spectrometer (XRS)," in *Advances in Cryogenic Engineering*, R. W. Fast, Ed. Boston, MA: Springer US, 1991, pp. 1349–1357.
12. I. S. Cooper, *Cryogenic Surgery – Introduction*. NY: Medical Examination Publishing Co., Inc., New York, 1971.
13. B. Rubinsky, "Cryosurgery," *Annu. Rev. Biomed. Eng.*, vol. 2, no. 1, pp. 157–187, 2000.
14. H. M. Skye, K. L. Passow, G. F. Nellis, and S. A. Klein, "Empirically tuned model for a precooled MGJT cryoprobe," *Cryogenics (Guildf).*, vol. 52, no. 11, pp. 590–603, 2012.
15. R. S. Dondapati and V. V. Rao, "Pressure drop and heat transfer analysis of long length internally cooled HTS cables," *IEEE Trans. Appl. Supercond.*, vol. 23, no. 3, p. 5400604, 2013.
16. H. Ding, Y. Wu, Q. Lu, Y. Wang, J. Zheng, and P. Xu, "A modified stress-strain relation for austenitic stainless steels at cryogenic temperatures," *Cryogenics (Guildf).*, vol. 101, pp. 89–100, 2019.
17. D. Raja Sekhar and V. V. Rao, "Three dimensional CFD analysis of cable-in-conduit conductors (CICCs) using porous medium approach," *Cryogenics (Guildf).*, vol. 54, pp. 20–29, 2013.
18. S. Spagna, J. Diederichs, and R. E. Sager, "Cryocooled refrigeration for sensitive measuring instrumentation," *Phys. B Condens. Matter*, vol. 280, no. 1, pp. 483–484, 2000.
19. E. W. Lemmon, M. L. Huber, and M. O. Mclinden, "NIST reference Fluid thermodynamic and transport properties – REFPROP," National Institute of Standards and Technology, Boulder, CO, 2010.
20. R. Radebaugh, E. D. Marquardt, J. Gary, and A. O'Gallagher, "Regenerator behavior with heat input or removal at intermediate temperatures," in *Cryocoolers 11*, R. G. Ross, Ed. Boston, MA: Springer US, 2002, pp. 409–418.
21. Z. Melhem, *High Temperature Superconductors (HTS) for Energy Applications*. Woodhead Publishing, Cambridge, United Kingdom, 2011.
22. G. F. Weston, *Ultrahigh Vacuum Practice*. London/Boston, MA: Butterworths, 1985.
23. A. Roth, *Vacuum Technology*. Amsterdam: Elsevier Science B.V., 1990.

24. R. Boetes and C. J. Hoogendoorn, *Heat Transfer in Polyurethane Foams for Cold Insulation. In Heat and Mass Transfer in Refrigeration and Cryogenics.* New York: Hemisphere Publishing Corp., 1987.

25. R. F. Barron, *Cryogenic Heat Transfer,* 1st ed. Taylor & Francis, Boca Raton, Florida, 1999.

26. H. R. G., "Development of urethane foams for LNG insulation," *Advances in Cryogenic Engineering;* Vol. 20; pp. 338–342, 1975.

27. R. H. Kropschot and R. W. Burgess, "Perlite for cryogenic insulation," in Proceedings of the 1962 Cryogenic Engineering Conference, ed. K.D. Timmerhaus, University of California, Los Angeles, California, pp. 425–436, 1963.

28. J. Ekin and G. Zimmerman, "Experimental techniques for low-temperature measurements: Cryostat design, material properties, and superconductor critical-current testing," *Phys. Today,* vol. 60, pp. 67–68, 2007.

29. L. G. Rubin, "Cryogenic thermometry: A review of progress since 1982," *Cryogenics (Guildf).,* vol. 37, no. 7, pp. 341–356, 1997.

30. C. Minas and E. T. Laskaris, "Structural design and analysis of a cryogen-free open superconducting magnet for interventional MRI applications," *IEEE Trans. Appl. Supercond.,* vol. 5, no. 2, pp. 173–176, 1995.

31. T. A. Keim, T. E. Laskaris, J. A. Fealey, and P. A. Rios, "Design and manufacture of a 20 MVA superconducting generator," *IEEE Trans. Power Appar. Syst.,* vol. PAS-104, no. 6, pp. 1474–1483, 1985.

32. Y. Kim, T. Ki, H. Kim, S. Jeong, J. Kim, and J. Jung, "High temperature superconducting motor cooled by on-board cryocooler," *IEEE Trans. Appl. Supercond.,* vol. 21, no. 3, pp. 2217–2220, 2011.

33. T. Zhang, K. Haran, E. T. Laskaris, and J. W. Bray, "Design and test of a simplified and reliable cryogenic system for high speed superconducting generator applications," *Cryogenics (Guildf).,* vol. 51, no. 7, pp. 380–383, 2011.

34. R. J. Laverman and B.-Y. Lai, "Method and apparatus for cooling high temperature superconductors with neon-nitrogen mixtures."

35. R. Radebaugh, "Refrigeration for superconductors," *Proc. IEEE,* vol. 92, no. 10, pp. 1719–1734, 2004.

36. K. M. Kim *et al.,* "Heat characteristic analysis of a conduction cooling toroidal-type SMES magnet," *Phys. C Supercond. Appl.,* vol. 470, no. 20, pp. 1711–1716, 2010.

37. J. A. Demko and W. V Hassenzahl, "Thermal management of long-length HTS cable systems," *IEEE Trans. Appl. Supercond.,* vol. 21, no. 2, pp. 957–960, 2011.

38. J. T. Kephart, B. K. Fitzpatrick, P. Ferrara, M. Pyryt, J. Pienkos, and E. M. Golda, "High temperature superconducting degaussing from feasibility study to fleet adoption," *IEEE Trans. Appl. Supercond.,* vol. 21, no. 3, pp. 2229–2232, 2011.

39. U. Fleck, D. Vogel, and B. Ziegler, "Cooling of HTS applications in the temperature range of 66 K to 80 K," *Adv. Cryogen. Eng.,* vol. 47, pp. 188–198, 2002.

40. L. Chen, Y. Tang, J. Shi, and Z. Sun, "Simulations and experimental analyses of the active superconducting fault current limiter," *Phys. C Supercond.,* vol. 459, pp. 27–32, 2007.

41. S. W. Schwenerly *et al.,* "Performance of a 1-MVA HTS demonstration transformer," *IEEE Trans. Appl. Supercond.,* vol. 9, no. 2, pp. 680–684, 1999.

42. M. Iwakuma *et al.,* "Development of a 3 phi-66/6.9 kV-2 MVA REBCO superconducting transformer," *IEEE Trans. Appl. Supercond.,* vol. 25, no. 3, pp. 1–6, 2015.

43. N. D. Glasson, M. P. Staines, Z. Jiang, and N. S. Allpress, "Verification testing for a 1 MVA 3-phase demonstration transformer using 2G-HTS Roebel cable," *IEEE Trans. Appl. Supercond.*, no. 11, p. 5500206, 2013.

44. H.-M. Chang, Y. S. Choi, S. W. Van Sciver, and K. D. Choi, "Cryogenic cooling system of HTS transformers by natural convection of subcooled liquid nitrogen," *Cryogenics (Guildf).*, vol. 43, no. 10, pp. 589–596, 2003.

45. Y. S. Choi, "Cryogenic cooling system by natural convection of subcooled liquid nitrogen for HTS transformers," Retrieved from http//purl.flvc.org/fsu/fd/FSU_migr_etd-3784, 2004.

46. I. Sauers, R. James, A. Ellis, E. Tuncer, G. Polizos, and M. Pace, "Effect of bubbles on liquid nitrogen breakdown in plane-plane electrode geometry from 100–250 kPa," *IEEE Trans. Appl. Supercond.*, vol. 21, no. 3, pp. 1892–1895, 2011.

47. S. R. Kim, J. Han, W. S. Kim, M. J. Park, S. W. Lee, and K. D. Choi, "Design of the cryogenic system for 100 MVA HTS transformer," *IEEE Trans. Appl. Supercond.*, vol. 17, no. 2, pp. 1935–1938, 2007.

48. C. Gondrand, "Air liquide cryogenic solutions for HTS refrigeration – focus on Turbo-Brayton," *Int. Cryog. Eng. Conf.*, 2014.

49. H. Nakagome, T. Okamura, and T. Usami, "Research on improvement in the efficiency of the GM refrigerator," *Int. Cryocooler Conf.*, pp. 187–193, 2007.

3

Superconducting Generators/Motors

Raja Sekhar Dondapati

CONTENTS

3.1 Introduction

Based on the report of the International Energy Agency (IEA), motors account for 43–46% of global energy consumption. Moreover, large motors (>375 kW output power) account for ~23% of this consumption. The world generated nearly 20 trillion kWh of electrical energy in 2009, and the means of generation was by mechanical-to-electric conversion by rotating generators. Comparing such power generation capacity to other mechanical-to-electrical conversion systems, photovoltaic system supplies about 20 billion kWh, or ~0.1%. Large electrical generators (>50 MW) dominate the global power generation. Given the size of the markets for new units' replacement and retrofitting, motors and generators are the prime targets for superconducting versions. This is due to the fact that superconductors can carry much larger currents compared to that of copper without ohmic losses. Hence, such machines have a higher electrical efficiency, with smaller size and lesser weight for the same power. Despite such potential benefits, there are no commercial superconducting motors or generators due to economic issues. Hence, in this chapter, descriptions of these machines are presented along with the available prototypes and the existing challenges which hinder the commercialization of such devices.

According to the US Department of Energy (DoE), motors account for 70% of all energy consumed by the domestic manufacturing sector. Large electric motors (>1000 horsepower) consume ~55% of the total energy generated in America. Moreover, 70% of these motors are suited to utilize high-temperature superconducting (HTS) technology. With suitable minor exceptions, nearly all cruise ships are built with electrical propulsion, thus adapting marine motors as their primary motive power.

Rotating machines (motors and generators) employ copper windings on rotor and stator. When current flows in these windings, resistive losses due to *Joule heating* takes place, leading to a significant waste in energy. Hence, any improvement in efficiency would cause huge savings in energy usage. Moreover, the utility of superconductors in offering zero resistance at a cryogenic environment can be utilized for reducing electrical energy loss, along with reduction in size and weight of power components and machinery.

The discovery of high-temperature superconductors in 1986 led to the development of a variety of rotating machines, employing superconducting windings. These HTS conductors operate at higher temperatures (between 25 K and 77 K), thus simplifying refrigeration cooling systems. This chapter describes the topology of alternate current (AC) rotating machines, along with design and manufacturing issues. Finally, machine analysis is presented for enabling simulations of these machines in an electric grid.

HTS motors are ideal to be employed in pumps, fans, compressors and belt drives deployed in industries, particularly those required for continuous operation. Further, they are suitable for large process industries, such as paper processing, steel milling, oil and chemical. Table 3.1 shows the various models of the superconducting rotating device built using superconducting technology.

3.2 Principles of Superconducting Motors and Generators

Electrical motors and generators are the most explored topics in electrical engineering. In this section, the principles and characteristics pertaining to superconducting machines will be discussed. Moreover, this chapter is restricted to electromagnetic (EM) motors and generators, due to their relevance to superconducting materials.

There are two popular types of electric rotating machines, namely, synchronous and induction. A synchronous rotating machine has two windings: DC winding located on the rotor and AC winding located in the stator. Such a configuration is most commonly available, and the locations of the two windings are interchangeable. An induction motor consists of a squirrel cage or a three-phase wound winding in the rotor. The rotor winding carries AC current at slip frequency, and the rotor frequency is equal to the line frequency when the slip is 1 (rotor stationary) and the slip frequency when the motor is operated at its nominal speed. For low ratings (<500 horsepower), induction motors are very popular; however, for large-scale applications in industries or ships, synchronous motors are preferred. Moreover,

TABLE 3.1

Superconducting Motor Design

Superconductor	Coolant	Poles	Power/rpm/Frequency/Torque	References
REBCO	Gaseous helium	6	7.5 kW/360 rpm	[1]
		2	20 kW/600 rpm/10 Hz	[2]
			500 kW/1800 rpm	[3]
		4	400 kW/3600 rpm/120 Hz	[4]
YBCO	Gaseous helium	6	7.5 kW/360 rpm/18 Hz/199 N-m	[5]
	Liquid nitrogen/gaseous helium	8	15 kW/360 rpm/24 Hz/398 N-m	[6]
YBCO			10.5 Hz	[7]
2G superconducting tape		2	7.5 W/5 Hz	[8]
2G superconducting wire	Liquid nitrogen	10	400 kW/250 rpm/20.83 Hz/15000 Nm	[9]
Bi-2223		8	164 kW/2700 rpm/585 N-m	[10]
	Neon gas		17000 kW/200 rpm	[11]
	Liquid nitrogen	2, 4	1000 kW/3600 rpm/1800 rpm/60 Hz	[12]
	Liquid helium (HTS field winding), liquid neon (field winding)	2	125 kW/11700 rpm	[13]
	Liquid neon (HTS field coil), water		1000 kW/3600 rpm	[14]
	Liquid nitrogen	4	2.7 kW	[15]
NbTi		4	7457 kW/300 rpm	[16]
	Liquid nitrogen/liquid helium			[17]
		8	150 kW/400 rpm/26.7 Hz	[18]
MgB$_2$		4	1000 kW/12000 rpm/800 N-m	[19]
		10	400 kW/250 rpm/20.83 Hz/19000 Nm	[20]
		2	11 kW/0.01–400 Hz	[21]
	Liquid nitrogen	10	400 kW/250 rpm/20.83 Hz	[22]

for induction motors, windings (both stator and rotor) experience AC current and are therefore not good candidates for superconducting windings. It should be kept in mind that superconductor losses are negligible under DC operating conditions. However, losses in the AC environment are large and economically difficult to remove. The involvement of high cost of superconductors and cooling systems limits the primary applications of motors to large sizes (>1000 horsepower) for fan drives and pumps in industrial and utility markets. Figure 3.1 shows the schematic representation of a typical AC synchronous machine. A magnetic field is created due to the rotating HTS field in the copper armature winding. Compared to a conventional motor, the magnitude of this field is typically twice. An air-core (non-magnetic) construction on the rotor and non-metallic teeth in the stator are present, enabling the air gap field to be increased without core loss and saturation problem, which are inherent in laminated iron stator and rotor cores. The copper armature winding lies outside the air gap, whereas in some applications, it is embedded in the non-metallic teeth to provide mechanical support. Under steady-state conditions, the rotor spins in sync with the rotating field which is created by the three-phase armature currents. Under a transient source or load, the rotor moves with respect to the field created by the armature current, and it experiences AC field harmonics. An electromagnetic shield is located between the HTS coils and the stator winding which shields the HTS field winding from these AC fields. A room temperature (warm) electromagnetic shield is positioned at the outermost surface of the rotor. Thermally insulated space (vacuum) is present inside the warm shield, which surrounds the rotor cryostat. The cold electromagnetic shield is positioned on the inner surface of this vacuum space and is highly conductive near the operating temperature of superconducting coils. The warm electromagnetic shield directly transfers torque to the warm shaft and is designed

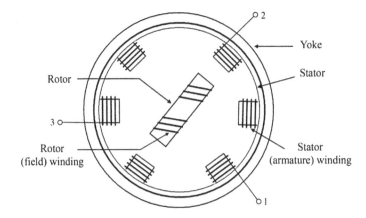

FIGURE 3.1
Configuration of a typical synchronous machine employing superconducting field winding on the rotor.

to be mechanically robust in order to withstand the large forces generated during faults. Moreover, it is designed to absorb the heat generated due to negative sequence currents and other harmonic currents generated by variable speed drive.

A refrigeration system is employed which uses forced circulation of helium gas (or other suitable gas) in a closed loop, which maintains the HTS field winding at a cryogenic temperature. Such gases are circulated through cooling channels inside the rotor. The closed cooling loop runs by the turning rotor to the externally located stationary refrigeration system. No magnetic iron teeth is employed by the copper stator winding and is typically designed with class F (maximum operating temperature: 155°C) insulation; however, it is operated at class B (maximum operating temperature: 130°C) insulation temperature. The stator employs the Litz conductor for the reduction of eddy-current losses, which is made of small diameter insulated and transported wire strands. The stator winding bore surface experiences a high magnetic field, which would saturate the iron teeth of a conventional stator, and thus the superconductor motor armature does not employ iron teeth. Due to the absence of iron teeth in the winding region, the cooling and support of the stator coils require special attention.

The first and obvious design difference between superconducting and conventional motors and generators, which utilize aluminum and copper windings, is the need to cool the superconducting windings to cryogenic temperatures. Copper windings in conventional machines require air or water cooling. Low-temperature superconductor (LTS) machines operate at ~4 K, and high-temperature superconductors (HTS) are the ones operated at 20–40 K. LTS and HTS wires have higher transition temperatures (T_c). However, engineering safety margins and high magnetic fields usually limits the operating temperatures. Moreover, considerable restriction on the overall machine design is imposed due to the requirement to cryocool the superconducting windings. The machine design should protect the superconductor winding as well as the cryogen from excessive heating due to hot machine parts and ambient heat. Such a process involves the minimization of thermal conduction using vacuum insulation and by employing low emissivity coatings for reducing radiative heating on hot areas. However, both the requirements are difficult and expensive to implement, which could further lead to reliability issues. For example, due to material outgassing, vacuum insulations can be difficult to maintain for longer periods. Furthermore, cryogenic cooling leads to thermal contractions, which must be properly designed into the mechanics of the contraction, including the stresses generated due to dissimilar thermal expansion coefficients of different parts.

There are various issues which hinder the infiltration of superconducting motors in commercial sectors, such as reliability and cost of the cryogenic subsystem. Another important consideration for superconducting motors and generators is the zero resistance offered by superconductors when under DC (direct current) operational mode. The flow of AC leads to losses in all superconductors; however, such losses can be reduced by making fine wire filaments that are twisted and transposed. However, this has not been accomplished for HTS. Moreover, the cost of cryo-refrigeration at 4 K makes it difficult to tolerate such losses in LTS, even in their best AC wire architectures. Such an effect limits the application of superconductors in field winding, which is substantially DC, while the AC armature windings remain copper. However, the exciter ramps the field during motor or generator operation to different field levels, in order to accommodate load or grid changes. Such situation causes losses in the superconducting field, which must be known and accommodated by the cryogenic system. Though, few designs and prototypes with superconducting armatures are available, the involved AC losses are quite high. Another important factor affecting the design and construction is the effect of AC magnetic fields on superconductors. Just as AC currents result in the heating of the superconductor, the same is true for impinging AC magnetic fields. Such fields are created by armature reaction, machine motions and equipment connected to the armature. Such conditions require the construction of a field damper, or EM shield around the superconducting field.

Further, the characteristic affecting a superconducting machine design is the potential loss of superconductivity in the superconductor winding in a phenomenon called *quench*. When the critical parameters of a superconducting wire, i.e., critical temperature (T_c), critical current (J_c) and critical external magnetic field (H_c), are exceeded, quench occurs. Superconductors are poor electrical conductors under *normal state*; hence, they will immediately experience large ohmic heating under quench conditions. If such conditions are not accommodated adequately, it will damage the machine by destroying the superconducting wire and insulation. Within the design of LTS, quench is commonly dealt by transferring the current into conductive materials surrounding the superconductor (for example, copper, often referred as stabilizer). Moreover, the propagation speed of a quench normal-zone is high in LTS, due to its low heat capacity at 4 K, which aids in minimizing the damage to heat localization. For HTS, the propagation of normal-zone is slow, making the protection much harder by tending to localize the heating. Further, it is also harder to produce a quench since larger energy transients are required for raising the temperature above the T_c.

The complete quench protection for the HTS system has been a subject of active research [23, 24]. The protection from quench results in the addition of software, hardware (e.g., temperature sensors) and control electronics to the design, which lead to the overall complexity compared to conventional copper designs. Further, design and construction differences arise due to the difference in properties of superconductors from conventional copper. While copper is very ductile and malleable, LTS material such as niobium titanium (NbTi) is ductile, but not malleable. This leads to the practice of attaching reinforcing materials to superconducting wires. Moreover, strain tolerance and bend radii of superconducting wires are considerably less than copper, which leads to the precaution arising during the design and construction of superconducting wires from all forces under normal operational and fault conditions. Other design parameters include the number of electrical phase and slots for the armature winding, rotor dynamics, pole count for the field winding, cooling of conventional components, insulation, bearing, torque transfer and vibration [25].

3.3 Types of Superconducting Motors and Generators

Several types of motors and generators have been explored for their applicability and benefits of superconductors in their construction. Figure 3.2 shows the classification of motors. The primary realization for such integration is that the motor or generator must be "large" due to the aforementioned cryogenic system, which must be added to the construction. Now, the quantification of large is to be established, which requires the knowledge of the details of the cryogenic machine design and machine cost versus conventional competitors. Researchers often use 1000 horsepower as the lower limit for motors and generators, respectively. A very common machine design utilized for motors and generators is the AC synchronous design. As the name describes, these machines rotate synchronously with the frequency (typically 50 Hz or 60 Hz) connected to the armature. The field winding is located on the rotor, whereas the armature is on the stator and is normally made of copper. Moreover, it is possible to rotate the armature and leave the field stationary, thus achieving the benefit of a stationary superconducting field [26]. In such cases, the "stator" and the "rotor" lose their traditional meaning. The AC synchronous design allows two distinct types: iron core and air core.

Iron core design follows the design philosophy of conventional motor and generator by using soft magnetic materials, in order to concentrate and direct the magnetic flux in the machine to the most advantageous locations, to produce optimal voltage in the armature windings and reduce the eddy-current heating in other metal parts of the machine. However, superconducting fields can be produced far in excess of the saturation magnetization of known soft magnetic materials. In such cases, the flux could not be contained by magnetic core materials and direct the flux, it rather acts as a source of eddy-current heating. In such cases, the construction comprising of air-core is the best approach. Figure 3.3 shows the schematic of a typical air-core superconducting generator. The magnetic iron is replaced with non-magnetic structural materials, which hold the armature and the field coils. Such a construction increases the air gap of the machine, resulting in a reduction in size, weight and synchronous reactance.

Another type of construction is called "trapped-field machines". In this configuration, a superconducting material, usually HTS, is utilized for trapping of

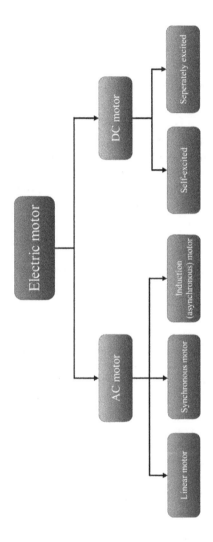

FIGURE 3.2
Classification of motors.

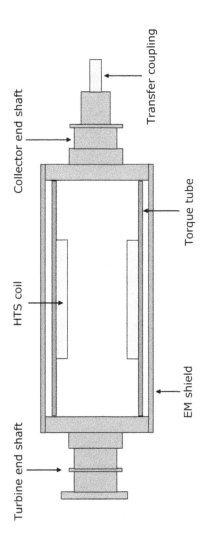

FIGURE 3.3
Schematic of an air-core superconducting generator.

magnetic flux, by cooling it in a magnetic field. Hence, larger fields compared to that of conventional permanent magnets, which are typically around 1 T or less, can thereby be produced. Thus, superconducting magnets can then be utilized as a field magnet in motors or generators [27, 28]. Hence, such machines can be thought as superconducting magnetic machines.

An issue related to such machines is that only as long as the superconducting magnet is kept cold, their field magnetism is permanent. If they warm up above their transition temperature, the superconducting magnets must be remagnetized. Such a condition leads to complex subsystems of coils and power supplies in order to remagnetize the machine field after a fault. Further classifications of motors and generators are based on two types: synchronous (operating in lockstep with the line frequency) and asynchronous. The most popular type of asynchronous generator and motor is the induction machine. In such machines, field current is induced into the closed-circuit field windings on the rotor, often shaped like a *squirrel cage*, by AC currents in the armature windings and the rotation of the field. The synchronous rotation of the rotor is called the slip frequency, with respect to the line frequency. Such an AC field current makes such a design not a good choice due to AC losses. The reluctance motors or generators use variations in the magnetic properties of the rotor. To create such situations, the salient poles on the rotor can be utilized, and superconducting materials in bulk form can be used to enhance this effect, due to the diamagnetic nature of superconductors. Diamagnetism in superconductors is the property which causes the rejection of magnetic flux and has a negative magnetic susceptibility.

As with conventional machines, both axial and radial magnetic flux designs for superconducting motors and generators are possible. Due to the ease in the design for construction of concentric cylinders for the stator and rotor, the more common deign is radial. The axial design favors either a stationary or rotating disc on a shaft next to each other. Such schemes have been employed for various superconductor designs [29, 30]. Due to the additional complication to take cryogen on and off the rotor, design approaches that allow the superconductor field to remain stationary have gained popularity.

A machine which allows such design is the AC homopolar inductor alternator [30–32]. Such machines utilize a stationary superconducting field, attached to the stator in order to induce magnetism in a high permeability rotor. In order to produce the armature voltage, the variable reluctance paths apply a time-varying magnetic flux to the armature winding. Another advantage of this design is that it allows solid rotors without coils, which are helpful in achieving high-speed operations. However, due to saturated magnetization of the rotor material, the magnetic field flux magnitudes are limited, and the design utilizes this flux less effectively (being homopolar) compared to that of a more traditional dipolar machine. In a dipolar machine, due to the rotation of field dipole, the flux magnitude goes through zero changes to its opposite directional sign and passes through the armature. In the case of homopolar design, the oscillation of flux takes place from a high to a low value, without changing to the opposite sign. Another design with stationary superconducting coils is the direct current (DC) homopolar machine. The DC machine obtains its torque from the Lorentz force between the axial current traversing the rotor and the radial magnetic field produced by the superconductor. Their advantages include the ease in speed/power control and quietness in operation. The disadvantages are the inherently high-current, low-voltage machines, and these currents must be taken on and off the rotor with brushes, which pose a reliability issue. Such machines have been designed and built [33, 34].

It must be noted here that the other useful components could be constructed using superconducting materials. For example, machine bearings may be constructed from superconducting bulk magnets, primarily HTS, for the higher operating temperature [35]. Such bearings give lower frictional losses and thus a longer life. Moreover, the superconducting versions have the advantage of being inherently stable, unlike conventional magnetic bearings, which are made stable through electronic control. HTS current leads may also be useful when employed in such machines for power connections to the superconducting windings [36]. The advantage of such leads is the lower heat transmission to the superconducting coils due to the low thermal conductivity of HTS. Furthermore, it is possible that the construction of machine exciters may become cryogenic with some superconducting parts, in order to gain efficiency [37]. However, none of the aforementioned components would likely be a superconductor in a conventional non-superconducting machine, due to the cost of supplying cryo-refrigeration. In a superconducting machine, which is integrated with a cryogenic subsystem, their use is more likely to be cost-effective.

3.4 Design Analysis and Parameter Calculation

With respect to the high cost of HTS and the complications involved during the manufacturing of stator with non-metallic teeth, alternate configurations of the synchronous machines have been considered. Several configurations are possible; however, four key options have been selected [38] and are shown in Figure 3.4. The details of the four options are described below:

(a) air-core stator and rotor. Both stator and rotor employ air core, i.e., non-magnetic iron core is employed in order to carry the magnetic flux. However, an iron core encircles the stator windings for shielding the space in the vicinity of the machine.

(b) air-core stator and rotor with iron yoke on the rotor. This is similar to (a); however, a solid iron yoke is employed in the bore of the field winding for the reduction in reluctance path between adjacent HTS field poles.

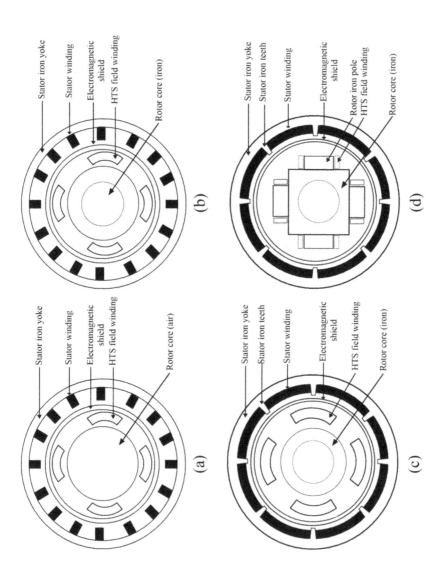

FIGURE 3.4
(a–d) Four possible configurations adapted for an HTS rotating machine.

(c) air-core rotor (with iron yoke on the rotor) and conventional iron core stator. This is similar to (b), except that the stator is of a conventional type, where coils are housed in the slots cut into the laminated iron core.

(d) iron core rotor and conventional iron core stator. This is similar to (c), except that the rotor has poles made of solid magnetic steel and HTS coils are installed around the poles.

It is possible to operate the poles and iron rotor core at cryogenic temperature. However, at cryogenic temperatures, the conventional iron becomes brittle. For minimizing the complexity and cost, the rotor iron in (b–d) is conventional magnetic iron and is operated at an ambient temperature. Several machines have been successfully built and tested using these options by various manufacturers.

The process of designing a machine with a non-linear iron in the magnetic circuit is iterative. The key steps involved in providing insight into the design process are mentioned below:

1. Define the machine cross section by selecting guessed values for HTS winding radial thickness and stator details.
2. Calculate self-inductance and mutual inductance of the winding with other windings using excitation levels, which does not saturate the iron components.
3. Excite field windings with a guessed excitation level.
4. Select an operating current for the selected HTS wire using the characteristics supplied (I_c as a function of field and temperature) by the manufacturer. The ratio between the selected operating current (I_o) and I_c should follow the relation below.

$$\frac{I_o}{I_c} = \frac{2}{3} \tag{3.1}$$

5. Select the stator coil conduction cross section suitable for the desired stator voltage and cooling approach and stator operating current I_a.
6. Excite armature phase coils (all other stator and rotor coils carry no current) and calculate self-inductance (L_a). The synchronous reactance is given by Equation (3.2),

$$x_d = 1.5\omega L_a \tag{3.2}$$

7. Calculate the open-circuit induced voltage (E_o) in the stator winding using the mutual inductance (L_{fa}) calculated in Step (2) and the field winding operating current (I_{ft}) selected in Step (4) with the following equation.

$$E_o = \omega L_{fa} \frac{I_{ft}}{\sqrt{2}} \tag{3.3}$$

8. Calculate terminal voltage of the generator using the selected stator current (I_a) in Step (5) as follows.

$$\text{Load angle } \delta, \sin(\delta) = I_a \frac{x_d \cos(\varphi) - r_a \sin(\varphi)}{E_o} \tag{3.4}$$

$$\text{Terminal voltage } V_t = E_o \cos(\delta) - I_a x_d \sin(\varphi) - I_a x_a \cos(\varphi) \tag{3.5}$$

where,
   φ = power factor angle
   x_d = synchronous reactance
   r_a = armature resistance

$$\text{Power output } = 3 V_t I_a \cos(\varphi) \tag{3.6}$$

9. If the output power (P_o) calculated in Step (7) is not correct, then go to Step (1), make necessary adjustments and repeat the calculations.

3.5 Summary and Future Trends

From the various successful prototypes of superconducting motors and generators, it is apparent that the technical challenges existing during its construction have been addressed. Such a condition raises the question: Why there are no commercial superconducting motors or generators at present? The answer lies in the economic point of view [39]. Such new machines must compete with the conventional designs, which demonstrate cost benefits. The extra cost is incorporated due to the superconducting wire and the operating and maintenance expenses of the cryo-refrigeration subsystem. What can be done in order to eliminate such economic issues? The cost of cryo-refrigeration can be reduced by the selection of superconductors having higher T_c. Further, it would also be helpful to operate in the temperature range of liquid nitrogen (77 K). However, such superconductors would not suffice in a current-carrying capacity at this temperature. Therefore, improvement in

HTS wire performance is an important aspect; however, the more significant aspect is the HTS wire price reduction.

There is no doubt regarding the various performance benefits superconducting motors and generators offer. Such benefits can be realized technically; however, the cost of incorporating superconductors can also be less. When the condition of cost reduction is achieved, then the future commercialization of superconducting machines is demonstrated.

References

1. M. Iwakuma *et al.*, "Production and Test of a REBCO Superconducting Synchronous Motor," *IEEE Trans. Appl. Supercond.*, vol. 19, no. 3, pp. 1648–1651, 2009.
2. K. Tamura *et al.*, "Study on the Optimum Arrangement of the Field Winding for a 20-kW Fully Superconducting Motor," *IEEE Trans. Appl. Supercond.*, vol. 26, no. 4, pp. 1–5, 2016.
3. A. Kawagoe *et al.*, "Numerical Analyses on the Influences of Armature Winding Shape and Yoke Arrangements on Total Losses in Fully Superconducting Synchronous Motors Using REBCO Tapes," *IEEE Trans. Appl. Supercond.*, vol. 28, no. 4, pp. 1–4, 2018.
4. M. Iwakuma *et al.*, "Feasibility Study on a 400 kW–3600 rpm REBCO Fully Superconducting Motor," *IEEE Trans. Appl. Supercond.*, vol. 22, no. 3, p. 5201204, 2012.
5. M. Iwakuma *et al.*, "Development of a 7.5 kW YBCO Superconducting Synchronous Motor," *IEEE Trans. Appl. Supercond.*, vol. 18, no. 2, pp. 689–692, 2008.
6. M. Iwakuma *et al.*, "Development of a 15 kW Motor with a Fixed YBCO Superconducting Field Winding," *IEEE Trans. Appl. Supercond.*, vol. 17, no. 2, pp. 1607–1610, 2007.
7. J. Li, F. Yen, S. Zheng, S. Wang, and J. Wang, "Normal Force Analysis on a High Temperature Superconducting Linear Synchronous Motor," *IEEE Trans. Appl. Supercond.*, vol. 22, no. 3, p. 5200304, 2012.
8. J. Lee *et al.*, "Electrical Properties Analysis and Test Result of Windings for a Fully Superconducting 10 HP Homopolar Motor," *IEEE Trans. Appl. Supercond.*, vol. 22, no. 3, p. 5201405, 2012.
9. L. Li, J. Cao, B. Kou, Z. Han, Q. Chen, and A. Chen, "Design of the HTS Permanent Magnet Motor with Superconducting Armature Winding," *IEEE Trans. Appl. Supercond.*, vol. 22, no. 3, p. 5200704, 2012.
10. P. J. Masson and C. A. Luongo, "High Power Density Superconducting Motor for All-Electric Aircraft Propulsion," *IEEE Trans. Appl. Supercond.*, vol. 15, no. 2, pp. 2226–2229, 2005.
11. S. Baik, Y. Kwon, S. Park, and H. Kim, "Performance Analysis of a Superconducting Motor for Higher Efficiency Design," *IEEE Trans. Appl. Supercond.*, vol. 23, no. 3, p. 5202004, 2013.
12. S. K. Baik *et al.*, "Design Considerations for 1 MW Class HTS Synchronous Motor," *IEEE Trans. Appl. Supercond.*, vol. 15, no. 2, pp. 2202–2205, 2005.

13. D. J. Waltman and M. J. Superczynski, "High-Temperature Superconducting Magnet Motor Demonstration," *IEEE Trans. Appl. Supercond.*, vol. 5, no. 4, pp. 3532–3535, 1995.

14. S. Baik and G. Park, "Load Test Analysis of High-Temperature Superconducting Synchronous Motors," *IEEE Trans. Appl. Supercond.*, vol. 26, no. 4, pp. 1–4, 2016.

15. W.-S. Kim, S.-Y. Jung, H.-Y. Choi, H.-K. Jung, J. H. Kim, and S.-Y. Hahn, "Development of a Superconducting Linear Synchronous Motor," *IEEE Trans. Appl. Supercond.*, vol. 12, no. 1, pp. 842–845, 2002.

16. C. L. Goodzeit, R. B. Meinke, and M. J. Ball, "A Superconducting Induction Motor Using Double-Helix Dipole Coils," *IEEE Trans. Appl. Supercond.*, vol. 13, no. 2, pp. 2235–2238, 2003.

17. J. P. Xu, S. Iino, and A. Ishiyama, "Design and Characteristics Analysis of Superconducting Tubular Linear Induction Motors," *IEEE Trans. Appl. Supercond.*, vol. 7, no. 2, pp. 660–663, 1997.

18. P. Tixador, F. Simon, H. Daffix, and M. Deleglise, "150-kW Experimental Superconducting Permanent-Magnet Motor," *IEEE Trans. Appl. Supercond.*, vol. 9, no. 2, pp. 1205–1208, 1999.

19. C. D. Manolopoulos, M. F. Iacchetti, A. C. Smith, K. Berger, M. Husband, and P. Miller, "Stator Design and Performance of Superconducting Motors for Aerospace Electric Propulsion Systems," *IEEE Trans. Appl. Supercond.*, vol. 28, no. 4, pp. 1–5, 2018.

20. L. Li, J. Cao, B. Kou, S. Yang, D. Pan, and H. Zhu, "Design of Axial and Radial Flux HTS Permanent Magnet Synchronous Motor's Rotor," *IEEE Trans. Appl. Supercond.*, vol. 20, no. 3, pp. 1060–1062, 2010.

21. A. Ishiyama and K. Hayashi, "Design and Construction of a Superconducting Cylindrical Linear Induction Motor with AC Superconducting Primary Windings," *IEEE Trans. Appl. Supercond.*, vol. 9, no. 2, pp. 1217–1220, 1999.

22. L. Li, J. Cao, Z. Sun, and Q. Chen, "Magnetic Field Distribution Around Superconducting Coils in Ferromagnetic Environment," *IEEE Trans. Appl. Supercond.*, vol. 21, no. 3, pp. 1131–1135, 2011.

23. W. D. Markiewicz, "Protection of HTS Coils in the Limit of Zero Quench Propagation Velocity," *IEEE Trans. Appl. Supercond.*, vol. 18, no. 2, pp. 1333–1336, 2008.

24. K. Sivasubramaniam *et al.*, "Transient Capability of Superconducting Devices on Electric Power Systems," *IEEE Trans. Appl. Supercond.*, vol. 18, no. 3, pp. 1692–1697, 2008.

25. I. Boldea, *Variable Speed Generators and Synchronous Generators*. CRC Press, Boca Raton, Florida, 2005.

26. E. T. Laskaris and K. Sivasubramaniam, "Method and Apparatus for a Superconducting Generator Driven by a Wind Turbine," *US Pat. 7821164*, 10/26/2010, vol. 1, no. 19, 2008.

27. H. Matsuzaki *et al.*, "HTS Bulk Pole-Field Magnets Motor with a Multiple Rotor Cooled by Liquid Nitrogen," *IEEE Trans. Appl. Supercond.*, vol. 17, no. 2, pp. 1553–1556, 2007.

28. D. Zhou, M. Izumi, M. Miki, B. Felder, T. Ida, and M. Kitano, "An Overview of Rotating Machine Systems with High-Temperature Bulk Superconductors," *Supercond. Sci. Technol.*, vol. 25, no. 10, p. 103001, 2012.

29. T. Okazaki, H. Sugimoto, and T. Takeda, "Liquid Nitrogen Cooled HTS Motor for Ship Propulsion," in 2006 *IEEE Power Engineering Society General Meeting*, June 18–22, 2006, Montreal, vol. 6, pp. 1–6, 2006.

30. H. Sugimoto *et al.*, "Design of an Axial Flux Inductor Type Synchronous Motor with the Liquid Nitrogen Cooled Field and Armature HTS Windings," *IEEE Trans. Appl. Supercond.*, vol. 17, no. 2, pp. 1571–1574, 2007.

31. Y. K. Kwon *et al.*, "Status of HTS Motor Development in Korea," in *2007 IEEE Power Engineering Society General Meeting*, 24–28 June 2007, Tampa, FL, pp. 1–5, 2007.

32. K. Sivasubramaniam, E. T. Laskaris, M. R. Shah, J. W. Bray, and N. R. Garrigan, "High-Temperature Superconducting Homopolar Inductor Alternator for Marine Applications," *IEEE Trans. Appl. Supercond.*, vol. 18, no. 1, pp. 1–6, 2008.

33. R. Marshall, "3000 Horsepower Superconductive Field Acyclic Motor," *IEEE Trans. Magn.*, vol. 19, no. 3, pp. 876–879, 1983.

34. R. J. Thome, W. Creedon, M. Reed, E. Bowles, and K. Schaubel, "Homopolar Motor Technology Development," in *IEEE Power Engineering Society Summer Meeting*, 21–25 July 2002, Chicago, IL, vol. 1, pp. 260–264, 2002.

35. F. N. Werfel *et al.*, "Superconductor Bearings, Flywheels and Transportation," *Supercond. Sci. Technol.*, vol. 25, no. 1, p. 14007, 2011.

36. K. Tsuzuki *et al.*, "Study of Bulk Current Leads for an Axial Type of HTS Propulsion Motor," *J. Phys. Conf. Ser.*, vol. 234, no. 3, p. 32059, 2010.

37. J. W. Bray, L. J. Garces, and G. E. Co., "Cryogenic Exciter," *US Pat. 8,134,345*, 3/13/2012, 2012.

38. C. Lewis, *Wind Power Generation and Wind Turbine Design*, Ed. by W. Tong, WIT Press, Southampton, Boston, 2010.

39. R. Schiferl, A. Flory, W. C. Livoti, and S. D. Umans, "High Temperature Superconducting Synchronous Motors: Economic Issues for Industrial Applications," in *2006 Record of Conference Papers - IEEE Industry Applications Society 53rd Annual Petroleum and Chemical Industry Conference*, 11–15 September 2006, Philadelphia, Pennsylvania, pp. 1–9, 2006.

4

Superconducting Magnetic Energy Storage (SMES)

Gaurav Vyas and Raja Sekhar Dondapati

CONTENTS

4.1 Introduction to SMES

The energy storage market is growing rapidly with emerging technologies driven by smart grids, the utilization of renewable energy, the aim to meet carbon emission and electricity-driven vehicles. The necessity for conventional power system integration with smart grids is increasing drastically with an increase in the power demand. However, these smart grids are interconnected with renewable power generation systems that produce a discontinuous power output. Hence, there is a necessity of a system which charges when the excess power is produced and discharges when the power demand increases. Superconducting magnetic energy storage (SMES) is one system that stores direct electrical energy with high specific power density, infinite discharge and charge cycles and discharges the electrical power within milli-seconds with an energy conversion efficiency of above 95%. The SMES utilizes three fundamental principles such as zero resistivity of the superconductor which corresponds to zero resistive losses, diamagnetism which corresponds to magnetic field repulsion and direct energy storage in a magnetic field. These principles provide the potential merits for using the superconducting coil to store electrical energy efficiently.

The operation of SMES is completely different from the conventional storage technologies because the continuous flow of electrical current in the superconducting coil generates the stored energy. Further, SMES utilizes direct current (DC). Hence, only one conversion process of alternating current (AC) to DC is required, and therefore, the thermodynamic losses in the conversion process are inherent. Initially, research was carried out to develop large storage SMES for load leveling application in integration with pumped hydroelectric storage. Further, researchers identified a potential merit of the rapid discharge of electrical power. Due to this capability, the technology of SMES is employed in electrical power systems for system stability and to pulsate power during voltage sags, power outage, load sensitivity etc. The economic viability is higher for the SMES system in terms of bulk energy storage for smaller systems due to high energy storage and rapid discharge potential. SMES is termed an impulsive source of current than an electrical energy storage system. Hence, it is an alternative for the uninterrupted power supply flexible AC transmission systems (FACTS) and power transmission and distribution systems in electrical power networks.

4.1.1 Working of SMES

A short circuited superconducting coil of SMES system stores the magnetic field due to DC electrical current, as shown in Figure 4.1. This system works efficiently when the temperature of the superconductor used in a superconducting coil of SMES is maintained below the critical temperature (T_c) at high magnetic fields. Hence, cryogenic coolants such as helium and nitrogen are employed in the operation of SMES to retain the superconductivity of the superconductor. At cryogenic temperatures (approximately 4.5 K), ohmic losses do not exist in the coil because the electrical resistance offered by the coil is 0 Ω (except at joints). Hence, higher efficiency (>95%) can be achieved (Figure 4.1).

There are four major components in SMES:

- **Superconducting coil** is made by superconductors such as YBCO, BSCCO and MgB_2. The coil is placed in the vacuum sealed, thermally insulated cryogenic environment using a cryostat. It is rested over mechanical support, consisting of low thermal conductance. Current leads are used to provide the electrical connections between the superconducting coil and the room temperature circuit. A detailed description is given in Section 4.2.

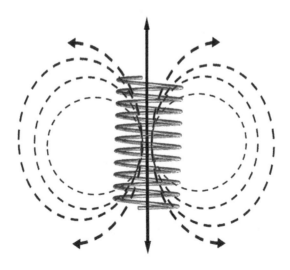

FIGURE 4.1
Concept of SMES (short circuited superconducting coil).

- **Cryogenic system** maintains the temperature of the superconductor below its critical temperature. The system is equipped with a cryostat, a cryocooler, compressors, vacuum pumps etc. A detailed description is given in Section 4.3.

- **Power conditioning system (PCS)** consists of power electronic devices, such as capacitors, transistors and inductors, that provide the interface between the superconducting coil (direct current) and the external grid or load (alternating current). The rated power of the SMES is determined by the power capacity of PCS. A detailed description is given in Section 4.4.

- **Control System** continuously monitors the magnetic protection, cryogenics, power electronics etc., with crucial parameters such as cryogenic coolant operation, coil strain, temperature and pressure. It is vital in the SMES because it establishes a link between the rated power demand from the external load or grid and the power flow from and to the superconducting coil. Information regarding the status of the superconducting coil and the dispatch feedback signal from the power grid is received and stored. The response of the SMES depends on the integration between the dispatch feedback and the charge level. It provides data to the operator on the status of the SMES system and maintains safety. Advanced SMES systems are integrated with the Internet to access control and remote observations (Figure 4.2).

4.1.2 Technical Aspects of SMES

Technical aspects that are to be considered while designing the SMES are:

i. **Power capacity of SMES**

The application in which the SMES is to be installed determines the power capacity of the SMES. For instance, applications such as load leveling, stability in the power system and power quality require different power capacities that depend on either the power rating of power conditioning system (PCS) or the maximum power that the coil can withstand ($P_{max} = I_{peak}V_{max}$).

ii. **Physical dimensions of SMES**

The volume of the SMES depends on the type of superconducting coil used, the type of cryogenic refrigeration system employed for cooling the superconducting coil and the subsystems used in the PCS. The superconducting coil mounted in the vacuum sealed cryostat is smaller in volume compared to that of the refrigeration system and the PCS.

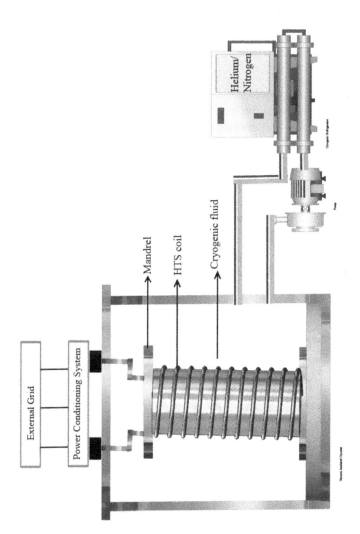

FIGURE 4.2
Liquid bath cooling of SMES.

iii. **Energy storage rating**

The energy storage ratings of the SMES depend on the integration of the application and location at which it is installed. It is the product of the discharge time of the electrical current to the load and the power capacity of the superconducting coil.

iv. **Efficiency of SMES**

In large-scale power applications, the overall efficiency of the SMES is as high as 95%. However, for the small-scale application, the efficiency is comparatively less than other energy storage systems. In SMES, the electrical energy is stored in the superconducting coils which exhibit absolutely zero losses during the operation. However, other subcomponents are not efficient than a superconducting coil due to the losses such as eddy current, hysteresis, conduction and radiation. The efficiency of the SMES depends on the energy stored in the superconducting coil. In PCS, few losses are encountered in the electronic subsystem during charging and discharging of the electric current, and therefore a change in the magnetic field can be noticed. Further, efficiency depends on the overall performance of the cryogenic refrigeration system.

4.1.3 Basic Principle of SMES

Superconducting magnetic energy storage (SMES) solely depends on the development of the superconductors that store energy in the form of a magnetic field. To analyze the SMES, a simplified series RL circuit is shown in Figure 4.3. The resistance offered by the coil and the internal resistance offered by the source result in the total resistance of the system. When the constant voltage source and the magnetic coil are connected, the electrical current varies as a function of time, a transient energy storage process in the magnetic coil will begin and the electrical current $i(t)$ in the circuit will

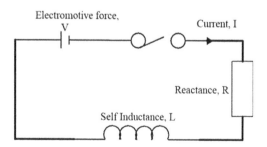

FIGURE 4.3
Simplified series RL circuit of SMES.

rise and fall during charging and discharging, respectively [1]. The transient process in the magnetic coil can be written as:

$$i(t) = \frac{1}{R}(V + e) \tag{4.1}$$

where, R is total resistance in the circuit, V is voltage of the power supply and e is induced electromotive force:

$$e = -L\frac{di(t)}{dt} \tag{4.2}$$

where, L is self-inductance of magnetic coil:

$$i(t) = \begin{cases} 0 \text{ at } t = 0; & \text{no magnetic field} \\ i_{max}; & \text{magnetic field} \end{cases} \tag{4.3}$$

Using Equation (4.3) in Equation (4.2), the transient process in the magnetic coil will be:

$$i(t) = i_{max}\left\{1 - e^{\left(-\frac{Rt}{L}\right)}\right\} \tag{4.4}$$

where, the maximum current is given by $i_{max} = \dfrac{V}{R}$ and the time constant is given by $t = \dfrac{L}{R}$.

Multiply $i(t)dt$ with the transient process to obtain the energy equation:

$$\frac{V i(t)dt - i^2(t)Rdt - Li(t)di(t)}{dt\,i(t)dt} = 0 \tag{4.5}$$

The energy supplied from the power supply should be equal to the energy stored in the form of the magnetic field and the energy losses in the magnetic coil.

$$dE_s = dE_m + dE_{losses} \tag{4.6}$$

The energy stored in the magnetic coil is given by:

$$dE_m = Li(t)di(t) \tag{4.7}$$

Solving Equation (4.7) to calculate the energy stored in the magnetic field at constant L:

$$E_m = L \int_{i=0}^{i_{max}} i(t) di(t) = L \left(\frac{i^2}{2} \right)_{i=0}^{i_{max}} = \frac{1}{2} L i_{max}^2 \qquad (4.8)$$

The energy stored in the magnetic coil depends on the number of turns (N) of the superconducting coil and the electrical current. From the expression of induced electromotive force (EMF) in the magnetic coil:

$$e = -Nd\phi dt \qquad (4.9)$$

$$E_m = \int_0^\phi N i(t) \phi = \int_0^B AlH \, dB = \frac{1}{2\mu} \iiint_{vol} BH \, dx dy dz \qquad (4.10)$$

There is an approximate linear relationship existing between H and B; Equation (4.10) can be simplified as:

$$E_m = \frac{Al\mu H^2}{2} = \frac{AlB^2}{2\mu} \qquad (4.11)$$

where, A is area of magnetic field, l is length of magnetic field, ϕ is magnetic flux, μ is permeability and B is induced magnetic field.

The magnetic coil made of a superconductor is preferred compared to that of a conventional conductor because when the superconducting coil is connected to a DC power supply, the electrical current in the coil increases thereby increasing the magnetic energy stored in the coil. Since superconductors offer zero resistance, the electrical energy supplied will be completely stored in the magnetic field, and continuous power input is not required.

4.2 Superconducting Coils

The superconducting coil is the heart of SMES, and the design of such a coil is crucial because it stores the maximum possible magnetic field that is required. For designing the coil efficiently, the factor that needs to be taken care of is the minimization of the volume of the superconductor for a desired energy storage, considering the mechanical stability and thermal stability with proper cryogenic cooling to ensure protection from quenching due to electromagnetic forces. Quenching of a superconducting coil may degrade the magnetic strength of the superconducting coil.

4.2.1 Superconducting Materials

The development of superconductors in large scale made a revolutionary change in superconducting power applications such as superconducting cables, SMES, superconducting motors, superconducting generators, superconducting transformers and superconducting fault current limiters. Superconductors are premium than normal conductors because they exhibit zero resistivity and handle higher current density at lower volumes. In SMES, multiple windings of superconductors are in the superconducting coil. The superconductors are manufactured either in wire or tape using different superconducting materials. The superconducting wires are manufactured using a thin superconducting material enclosed within aluminum, copper and silver alloy matrix to enhance the strength and to protect the wire from quenching. Parameters such as critical magnetic field (B_C), critical current density (J_C) and critical temperature (T_C) characterize the superconducting material. At cryogenic temperatures, there is a direct dependency within these three parameters to achieve a higher magnetic field and higher current densities as shown in Figure 4.4.

For instance, for SMES, higher magnetic fields are required to achieve higher storage capacities; consequently, the critical current density and the critical temperature are changed accordingly. The SMES systems are operated at a temperature of 4.5 K, and the critical temperatures of different superconductors are higher than 20 K (see Table 4.1). Similarly, in high-temperature superconductor (HTS) cable applications, higher current densities are required; consequently, the critical magnetic field and the critical temperature are changed accordingly. The HTS cables are operated at an operating range of 65–77 K, depending on the type of superconducting material used. The cryogenic coolants to be used for cooling the superconductors are

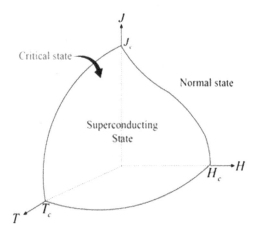

FIGURE 4.4
J-H-T curve of a superconductor.

TABLE 4.1

Classification of Superconductors Depending on Critical Temperatures [2]

Superconducting material	Superconductors (Critical temperature (K))	Cryogenic coolant (Boiling temperature (K) @ 1 atm)	Critical temperature of coolant (K)	Critical Pressure of coolant (K)	Latent heat of vaporization (kJ/kg)
LTS	NbTi (9.8)	Helium (4.2)	5.2	2.3	21
	Nb_3Al (18)	Hydrogen (20.3)	32.9	12.8	445
	Nb_3Sn (18.1)				
MTS	MgB_2 (39)	Hydrogen (20.3)	32.9	12.8	445
		Neon (27.1)	44.4	26.5	86
HTS	BSCCO-2212 (85)	Nitrogen (77.3)	126.3	34	199
	BSCCO-2223 (110)	Argon (87.302)	150.69	48.630	163
	YBCO (93)	Oxygen (90.2)	154.6	50.4	213

made up of different materials with respect to the critical temperature range as given in Table 4.1.

Superconductors are classified into two categories depending on the behavior of the superconductor in a magnetic field at a critical temperature and the critical temperature of the superconducting material. Based on the magnetic field, superconductors are either Type I or Type II. In Type I, superconductors act as perfect diamagnetics and repel the magnetic field when they are maintained below the critical temperature and are unable to sustain higher magnetic fields. In Type II, superconductors exhibit a mixed state around the critical temperature and can sustain higher magnetic fields without exhibiting the quench behavior. The superconducting materials used in the SMES project based on the energy stored are shown in Table 4.2. Based on the critical temperatures, superconductors are classified into low-temperature superconductors (LTS), medium-temperature superconductors (MTS) and high-temperature superconductors (HTS). The superconducting materials used in the SMES project based on the operating temperature are shown in Table 4.3.

Figure 4.5 shows the superconductors used in storing the energy in SMES. These superconductors are manufactured in the form of wires such as LTS, BSCCO and MgB_2 or as tapes with multi-layers, sandwiching the powdered superconductor over deposition layers and buffer layers such as (Re)BCO. The key issue in SMES is the selection of the superconductor based on the design and application. The optimal superconducting material should be selected, depending on parameters such as performance, cost, reliability, availability and manufacturing complexity.

4.2.2 Superconducting Coil Topologies

In SMES, different topologies are designed using various superconducting materials, depending on the availability of the superconductor, and the application and technical attributes of the SMES. In this section, designs of different SMES topologies such as solenoid, toroidal, four pole, racetrack, shielded and n-polygon topologies are presented.

4.2.2.1 Solenoid Topology of SMES

The solenoid topology of SMES is simple in construction. The dimensions of the solenoid SMES depend on the coil dimensions such as diameter, height and winding thickness. Aspect ratio (A_s) is defined as the ratio of the winding height of the coil to the diameter of the coil. The energy stored in the solenoid superconducting coil can be calculated using the aspect ratio to the maximum allowable magnetic field through the superconductor (B). Once the superconducting coil dimensions are predetermined based on the application, the design of the mandrel (structural steel) and the volume of the superconductor required are calculated. The critical current that the

TABLE 4.2

Superconducting Materials Used to Store Energy in SMES Projects around the World in Different Applications

Superconducting material	Energy (MJ)	Application	Country	Organization
NbTi	0.27	Power system stability	Japan	Tokyo Institute of Technology [3]
	1	UPS		National Nuclear Science Institute [4]
	2.9	Power system stability		Kyushu Electric Power Company [5]
	7.34	Voltage drop compensation		Chubu Electrical Power Company [6, 7]
	20			
	0.2	Power quality	Italy	Bologna University [8]
	2.6	Load sensitivity		Ansaldo Ricerche Spa [9]
	0.3	Voltage drop compensation	China	Tsinghua University [10]
	2	Impulse power source		Institute Of Electrical Engineering, CAS [11]
	1-5	Commercial application	USA	American Superconductor [12]
	30	Low frequency damping		Los Alamos Laboratory [13]
	100	Low frequency damping		University of Florida [14]
	3	UPS	Korea	Korean Electric Research Institute (KERI) [15]
Bi 2212	1	Voltage stability	Japan	Chubu [16]
	0.8	Impulse power source	France	DGA/CNRS [17]
NbTi, Bi 2212 (Hybrid SMES)	6.5	Experimental study	Japan	Toshiba Company [18]

(Continued)

TABLE 4.2 (CONTINUED)

Superconducting Materials Used to Store Energy in SMES Projects around the World in Different Applications

Superconducting material	Energy (MJ)	Application	Country	Organization
Bi 2223	0.01	Smart micro power grid (synchronization control)	China	University of Electronic Science and Technology of China and Innopower Corporation [19]
	0.03	Experimental study		Institute Of Electrical Engineering, CAS [20]
	0.035	Hubei province (Hydropower station)		Huazhong University of Science and Technology [21]
	1	Power quality		CAS [22]
	0.01	Voltage drop compensation	Korea	Changwon National University [19]
	0.6	Voltage and power stability		KERI [23, 24]
	2.5	Power stability		KERI [25]
	0.15	UPS	Germany	ACCEL Instruments Gmbh [26]
	0.0348	UPS	Poland	Superconducting Technology Laboratory [27, 28]
	0.0248	Voltage drop compensation	Australia	Wollongong University [29]
YBCO	2.5	Power quality	Korea	KERI
		DG Networks	USA	Superpower, ABB, BNL, TcSUH
	2400	Load fluctuation compensation	Japan	Chubu and Waseda University [30]

TABLE 4.3

Superconducting Materials Used at Different Temperature in SMES Projects [31]

Superconducting materials	Temperature (K)	Energy (MJ)	Country	Organization	Year
NbTi	4.2	0.0188	Germany	Forschungszentrum Karlsruhe	1995
		1	USA	Superconductivity Inc.	1988
		30	USA	Bonneville Power Authority	1982
BSCCO, NbTi	4.2	6.5	Japan	Toshiba	2007
BSCCO	4.2	1	China	Chinese Academy of Science	2008
	20	0.015	Germany	ACCEL Instruments	2003
		0.06	Korea	Seoul National University	2005
		0.0814	France	CNRS	2008
	25	0.005	USA	American Superconductor	1997
	64-77	0.0012	Israel	Bar-Ilan University	2003
BSCCO, YBCO	65-77	0.006	China	China Electric Power Research Institute	2010
YBCO	77	0.002	UK	University of Bath	2014

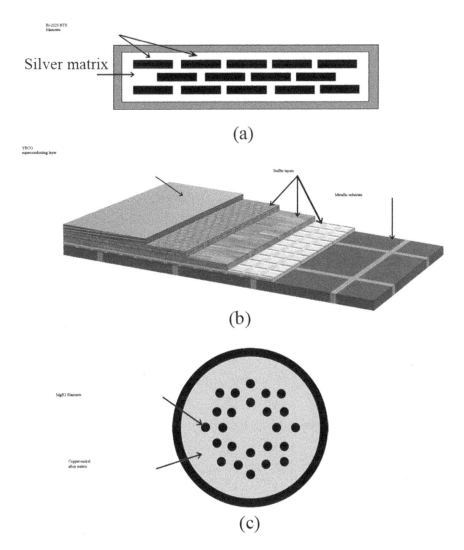

FIGURE 4.5
Superconductors used in SMES.

superconducting coil can withstand is set according to the mid-plane mag-netic field. When the electrical current flows through the superconducting coil, hoop stresses are developed that result in the radial Lorentz forces in the coil. Figure 4.6 shows the design of the solenoid SMES topology.

For a thin solenoid, the volume of the superconductor required is given by [32],

$$Q_{sc} = \frac{C_s(A_s)E^{0.667}}{B^{0.334}} \tag{4.12}$$

FIGURE 4.6
Solenoid type SMES topology.

where, $A_s = \dfrac{h}{2R}$ is the aspect ratio, s is solenoid, h and R are height and radius
of the solenoid.

The energy stored per unit volume of the superconductor is given by,

$$E_s = \frac{E_{sc}}{Q_{sc}} = \frac{C_s E^{0.334} B^{0.334}}{A_s} \qquad (4.13)$$

4.2.2.2 Toroidal Topology of SMES

The toroidal SMES is complex in geometry compared to the solenoid topology. However, the energy capacity of the toroidal SMES is higher than the
solenoid. The total energy stored in the superconducting windings can be
calculated similar to the thick-walled coil, because the magnetic field generated due to flow of electrical current is stored in the toroidal windings. In
the inner strut of the superconducting coil, maximum magnetic field will
occur. Further, the design of torus consists of discrete coils in large sections,
in order to minimize magnetic field losses. For large storage capacity, the
toroidal design requires buckling cylinders additionally to compensate the
net inward forces that result in large Lorentz forces on the windings of
the superconducting coil. The thickness of the buckling cylinder used in this

topology can be calculated from the hoop stresses that are developed in the cylinder. In toroidal topology, two configurations of the SMES are available, depending on the application, and are shown in Figure 4.7.

The inductance in the toroidal topology is given by [32],

$$L = \mu_0 R \left(1 - \left(1 - \left(\frac{r}{R} \right)^2 \right)^{0.5} \right) \tag{4.14}$$

where, r is minor radius and R is major radius

Energy stored in the toroidal topology is given by,

$$E_{SC} = 0.5 I_C^2 \, \mu_0 R \left(1 - \left(1 - \left(\frac{r}{R} \right)^2 \right)^{0.5} \right) \tag{4.15}$$

The maximum field in the toroidal topology is given by,

$$B_m = \frac{\mu_0 I_C}{2\pi R \left(1 - \left(\frac{r}{R} \right)^2 \right)} \tag{4.16}$$

The volume of the superconductor required for the coil is given by,

$$Q_{sc} = 2\pi a I_C^2 = 2\pi r I_C \tag{4.17}$$

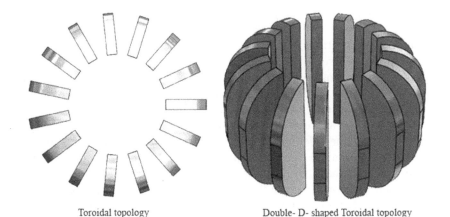

Toroidal topology   Double- D- shaped Toroidal topology

FIGURE 4.7
Toroidal topology of SMES.

$$Q_{sc} = \frac{2\pi r I_C^{1.334}}{I_C^{0.334}} = \frac{E^{0.667}}{B^{0.334}} \left[\frac{16\pi^2 \left(\dfrac{r}{R}\right)^3}{\left[\mu_0 \left(1-\left(\dfrac{r}{R}\right)\right)\left(1-\sqrt{1-\left(\dfrac{r}{R}\right)^2}\right)^2\right]^{0.33}} \right] \qquad (4.18)$$

Equation (4.7) can be rewritten as.

$$Q_{sc} = \frac{CP\left(\dfrac{r}{R}\right)E^{0.667}}{B^{0.334}} \qquad (4.19)$$

The energy stored per unit volume of the superconductor is given by,

$$E_s = \frac{E_{SC}}{Q_{SC}} \qquad (4.20)$$

Using Equation (4.15) and Equation (4.20), we get,

$$E_s = \left(E_{SC}B\right)^{0.334} \left[\mu_0\left(1-\left(\frac{r}{R}\right)\right)\left(1-\sqrt{1-\left(\frac{r}{R}\right)^2}\right)^2 16\pi^2\left(\frac{r}{R}\right)^3\right]^{0.334} \qquad (4.21)$$

4.2.2.3 Shielded Topology of SMES

SMES technology is advancing in small-scale power applications to compensate the fluctuations in the voltage drop and for stabilization of the power line. In such applications, the superconducting coil in the SMES encounters AC losses (high pulse losses) within a short span of charging and discharging due to rapid change in the magnetic fields. In order to reduce such losses, a shielded topology of SMES is designed that consists of the superconducting coil, shielded parallel with a normal coil, and as a result, the fluctuating current is shielded and constant electrical current flows through the superconducting coil. The shielded type SMES is designed in solenoid and toroidal topologies as shown in Figure 4.8. In a solenoid shielded type SMES, the shielding coil is placed in the space provided in the superconducting coil [33]. The field generated by the shielded coil may affect the magnetic field of the superconducting coil. For reducing the influence of the magnetic field, a double shielded solenoid type coil with high coaxial winding density is preferable. The advantage of the solenoid shielded SMES is that it is simple

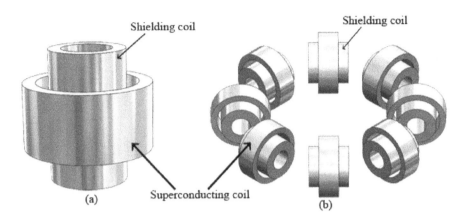

FIGURE 4.8
Shielded topology of SMES.

in terms of construction of the coil and the cryostat design with the refrigeration system. In toroidal shielded type SMES, the magnetic field developed in the superconducting coil is confined [34]. The fields due to fluctuating and stray components are small from the shielded coil on the superconducting coil. However, the design of toroidal shielded type SMES is complicated when it comes to arranging the shield in a torus.

The difference between the shielded topology and other topologies is the mutual inductance between the shielded and superconducting magnetic coils because it affects the forces acting on the superconducting coil. The correction in the mutual inductance depends on the electrical current flow in the shielding and superconducting coils.

The self-inductance in the shielding coil should be equal to the mutual inductance to satisfy the condition of shielding [33].

$$L_{Sh} = M \tag{4.22}$$

Maximum energy stored in the system is given by:

$$E = \frac{1}{2} L_{SC} I_{SC}^2 + M I_{SC} I_{Sh} + \frac{1}{2} L_{Sh} I_{Sh}^2 \tag{4.23}$$

where, L_{SC} is self-inductance in superconducting coil, L_{Sh} is self-inductance in shielding coil and M is mutual inductance of the system.

From Equation (4.23), the energy stored in the superconducting coil is given by:

$$E_{SC} = \frac{1}{2} L_{SC} I_{SC}^2 \tag{4.24}$$

In the shielding coil, at energy absorption state, the change in energy is given by:

$$\Delta E = MI_{SC}I_{Sh} + \frac{1}{2}L_{Sh}I_{Sh}^2 \qquad (4.25)$$

Equation (4.25) can be written as:

$$\Delta E = \lambda_{Sh}F_{Sh}^2\left(\frac{1}{2} + \frac{1}{K}\right) \qquad (4.26)$$

where, $K = \dfrac{I_{Sh}}{I_{SC}}$, and F_{Sh} is electromotive force (EMF) of the shielding coil.

The EMF of the superconducting coil is given by:

$$F_{SC} = \frac{\lambda_{Sh}}{mK_i}F_{Sh} \qquad (4.27)$$

The amount of superconductor and normal conductor required is given by:

$$Q_{SC} = 2\pi R_{SC}F_{SC} \qquad (4.28)$$

$$Q_{Sh} = 2\pi R_{Sh}F_{Sh} \qquad (4.29)$$

4.2.2.4 Other Configurations of Basic Topologies of SMES

In general, the solenoid and toroidal topologies are majorly used for energy storage in SMES applications. The major advantage of a solenoid is that it is simple in construction and can store higher energy; however, the strong drift in magnetic fields developed within the SMES can cause a harmful impact on the environment. Similar fields are observed in the modular toroidal coils that are designed with equally spaced pancake coils. However, such fields are available within the helical toroid where the superconductor is wound continuously. These are eco-friendly and can store huge amounts of magnetic fields.

To utilize the advantage of the solenoid and the modular toroid, different modifications in the geometries are researched to increase the energy storage capability of the superconducting coil, for a particular volume of the superconductor. The various topologies of the superconducting coils, for storing magnetic fields, include polygon, racetrack and four pole.

a) **n-polygon configuration**

Based on the availability of space, the unidirectional design of solenoids can be modified to form polygons by splitting the single solenoid into multiple solenoids with reduced dimensions that are placed either horizontal or vertical to each other [35, 36], as shown in Figure 4.9. These compact configurations are designed using the

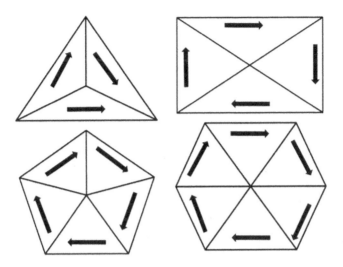

FIGURE 4.9
n-Polygon configuration of SMES.

same volume of the superconductor required for the unidirectional superconducting coil, with a relatively lower energy storage loss of approximately 5%.

b) **Racetrack coil configuration**

The racetrack consists of a pack of 24 modules of racetracks, placed one over the other as shown in Figure 4.10. These modules are placed similar to the modules of the helical toroid. The advantage with this configuration is that the case walls of the racetrack respond to the various Lorentz loads [37].

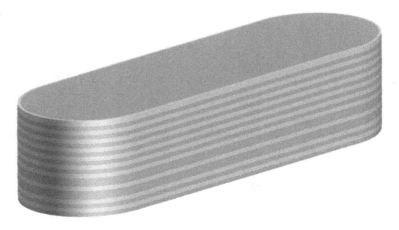

FIGURE 4.10
Racetrack coil configuration of SMES.

c) **Four pole configuration**

The four pole coils are employed due to compactness, less volume of the superconductor required and the economical and small stray field compared to the toroidal topology. Two configurations such as flat and long pole are investigated to decrease the stray magnetic field as shown in Figure 4.11. These coils are suitable for high-temperature superconductors such as BSCCO and YBCO and medium-temperature superconductors such as MgB_2 [38, 39]. The parameters that influence the magnetic fields are the inner radius of the coil, thickness of the coil, length of the coil, radius of the pole from the orientation center and the operating electrical current.

4.2.3 Generalized Thermo-Electrical Strategy for Designing a Superconducting Coil

The designing of a superconducting coil depends on the electric current flow in the coil and the joule heat generation due to flow of the electrical current [40].

Heat transport in SMES can be solved using conventional heat transfer equation:

$$\rho_g c_p \frac{\partial T}{\partial t} - \rho_g c_p u \cdot \nabla T = \nabla \cdot (k \nabla T) + Q \qquad (4.30)$$

where, ρ_g is density of coil configuration used, c_p is specific heat of coil, T is temperature, Q is heat source, k is thermal conductivity of coil u is velocity vector.

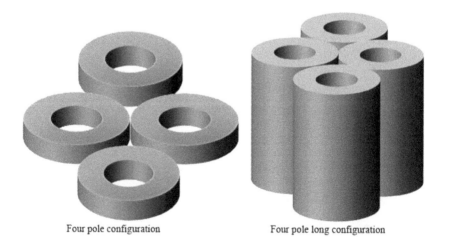

Four pole configuration   Four pole long configuration

FIGURE 4.11
Four pole configurations.

Since convection can be neglected in solids, Equation (4.30) can be written as:

$$\rho_g c_p \frac{\partial T}{\partial t} - \nabla \cdot (k \nabla T) = Q \tag{4.31}$$

From the joule heating and from the basis of superconductivity:

$$Q = E \cdot J \tag{4.32}$$

where, J = total current density and E = electric field strength; Equation (4.22) can be written as:

$$E = -\nabla V \tag{4.33}$$

where, V = electric potential
From E-J power law:

$$\frac{E}{E_0} = \left(\frac{J}{J_c} \right)^n \tag{4.34}$$

$$J = \left(\frac{1}{\rho} + \varepsilon \frac{\partial}{\partial t} \right) E + J_e \tag{4.35}$$

where, J_e = total current density, ε = electric permittivity and ρ = resistivity of the superconductor.

Further, the heat conducted due to the electric current flow and the joule heat generation from the HTS coil is dissipated to the cryostat.

Electrical resistivity can be calculated by:

$$\frac{T}{\rho k} = \text{constant} \tag{4.36}$$

$$-n(-k \nabla T) - q_0 = h(T_\infty - T_s) + e\sigma \left(T_{amb}^4 - T_s^4 \right) \tag{4.37}$$

where, n = unit vector normal to a boundary surface of the HTS coil, h = heat transfer coefficient of cryogenic fluid, q_0 = inward heat flux, T_∞ = bulk temperature of cryogenic fluid, T = surface temperature, e = emissivity of the coefficient of the coil, σ = Stefan–Boltzmann constant. The convection heat transfer coefficient depends on geometrical configuration of HTS coil, surface temperature and thermophysical properties of cryogenic coolant.

4.2.4 Thermal Stability of a Superconducting Coil

Superconductors are cryogenically cooled to retain superconductivity, the magnetic energy stored in the superconducting coil of SMES should be larger and is possible at higher electrical current flow. In such situations, if the superconductors exceed the critical operating conditions such as magnetic field, current density and temperature, the superconductors transform into normal conductors. The superconductor needs to be cooled below its critical temperature, because there is a chance of run-off and the conductors remain in the state of a normal conductor. Hence, thermal stability is required to analyze hot spots in the superconductor that lead to run-off. The hot spot temperature in the superconducting coil can be estimated using the thermal adiabatic equation.

Heat generated in the conductor due to flow of electric current is given by,

$$Q_g = I^2 R \tag{4.38}$$

$$q_g = \frac{Q_g}{\text{volume}} = \frac{I^2 R}{A*L} = \frac{I^2 \rho L}{A*L*A} \quad \left[\because R = \frac{\rho L}{A}\right] \tag{4.39}$$

$$q_g = \left(\frac{I}{A}\right)^2 \rho = J^2 \rho \tag{4.40}$$

$$q_g(t) = \rho(T) j(t)^2 \tag{4.41}$$

where, A is area of superconductor, C_p is specific heat of superconductor, ρ is resistivity of superconductor and L is length of superconductor.

From the energy balance,

Energy generated in the superconductor = Energy gained by the cryogenic coolant per unit time

$$q_g(t) = \rho(T) j(t)^2 = \dot{C}_p(T) \frac{dT}{dt} \tag{4.42}$$

Solving the above expression for current density,

$$j(t)^2 dt = \frac{C_p(T)}{\rho(T)} dT \tag{4.43}$$

Integrating the above equation for obtaining the maximum temperature,

$$\int_0^\infty j(t)^2 dt = \int_{T_0}^{T_{max}} \frac{C_p(T)}{\rho(T)} dT = f(T_{max}) \tag{4.44}$$

$$f(T_{max}) = J_0^2 \left[t_{det} + \frac{E_0}{I_0 V_{max}} \right] \tag{4.45}$$

$$f\left(T_{\max}\right) = J_0^2 \left[t_{\det} + \frac{E_0}{P_{\max}} \right] \quad \left[\because P_{\max} = I_0 V_{\max} \right] \tag{4.46}$$

where, R_d is discharge resistance, $t_{\det}$ is time of quench detection, E_0 is energy stored in superconducting coil, I_0 is current, J_0 is current density and $V_{\max}$ is maximum voltage across the SC magnet.

The function T is maximum if the ratio of $\dfrac{E_0}{P_{\max}}$ is low, which is favorable for the efficient operation of the SMES.

4.2.5 Mechanical Stability of a Superconducting Coil

The mechanical stability relates to the stresses that are developed in the superconducting coil during the flow of electrical current through it. The stresses generated in the coil, such as tensile, compressive and hoop stresses, affect the magnetic field stored in the coil, thereby decreasing the efficiency of the SMES. The mechanical stability of the structural mass with generalized force balancing system for the magnets can be calculated using the virial theorem [41]. This theorem is independent of the electrical current distribution, the configuration of the superconducting coil and the amount of energy stored in the superconducting coil.

In Figure 4.12, F_N is related to normal forces due to i_θ at the toroidal surface current, $i_\phi(\theta)$ represents dipole surface current results in normal and tangential forces and F_T corresponds to tangential forces. The total structural mass

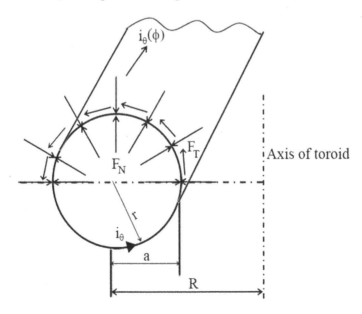

FIGURE 4.12
Distribution of magnetic force in the superconducting coil.

of the electromagnetic system with average unidirectional stressed structure $|\pm\sigma|$ and the density ρ is given by:

$$m_{st} = \frac{(1+2q_c)\rho_{st}E_s}{\sigma_{st}} \tag{4.47}$$

where, q_c is compression quality factor ($0 \geq q_c \leq 1$), $q_c = 0$ when the structure is under tension, σ_{st} is average design stress and E_s is energy stored in the superconductor.

If $q_c = 0$ then,

$$m_{st} = \frac{\rho_{st}E_s}{\sigma_{st}} \tag{4.48}$$

From the Clausius virial theorem:

$$m_{st} \geq \frac{\rho_{st}E_s}{\sigma_{st}} \tag{4.49}$$

$$m_t - m_c \geq \frac{\rho_{st}E_s}{\sigma_{st}} \tag{4.50}$$

where, m_t, m_c are mass in tension and mass in compression:

$$m_{st} \geq 2m_c \pm \frac{\rho_{st}E_s}{\sigma_{st}} \tag{4.51}$$

where, $m_{st} = m_t + m_c$

For instance, a toroidal configuration is considered, and the energy stored in the superconducting coil is given by:

$$E = \frac{1}{2}LI^2 \tag{4.52}$$

where, $L = \mu_0 R_T f(\beta)$ is inductance in the superconductor, R_T is major radius, β is aspect ratio $\left(\beta = \dfrac{a}{R}\right)$, a is characteristic length, I is electrical current and f is function of β.

If a shielding coil is used, the inductance relation between the uniform surface current and the uniform shielding current, which are perpendicular to each other, is given by:

$$f(B) = n_T^2\left\{1 - \left(1 - \beta^2\right)^{0.5}\right\} + n^2\left[\ln\frac{8}{\beta} - 2\right] \tag{4.53}$$

where, n_T is number of turns of toroidal windings and n_s is number of turns of shield windings

i. **Net radial forces acting on the superconductor**

From Figure 4.12, the forces F_N are normal to the toroidal windings and point outward and the tangential forces F_T are tangential to the surface and point inward. The net radial forces on the structures under compression or tension (depending on the magnitude of forces) are given by F_R and the mass bearing capacity in tension or compression.

$$m_R = \frac{2\pi\rho_{st}R_T T_{st}}{\sigma_{st}} \tag{4.54}$$

where, T_{st} is tension or compression in the structure wall.

$$T_{st} = -\frac{1}{4\pi}\frac{\partial L}{\partial R_T}I^2 \tag{4.55}$$

Solving Equation (4.55), using inductance

$$m_R = q_R \frac{\rho_{st}E_s}{\sigma_{st}} \tag{4.56}$$

where,

$$q_R = -\left(1 - \frac{\beta}{f}\frac{\delta f}{\delta\beta}\right) \tag{4.57}$$

If q_R is positive, the structure is under compression, and if q_R is negative, the structure is under tension.

ii. **Circumferential forces acting on superconductor**

If the shielding coil is considered, the tangential forces are neglected. From the force balance, the circumferential forces acting on the structure can be analyzed under compression or tension based on the magnitude. These can be analyzed using the virtual model as shown in Figure 4.13.

To calculate the structure, which can withstand the net magnetic forces that are normal to the surface (magnetic circumferential surface), the energy stored in the elemental surface of the structure is,

$$\Delta E_s = -\int T_{st}\partial S = \int T_{st}\left(\Delta rd\theta\right) \tag{4.58}$$

From Figure 4.13, it can be seen that

$$\Delta r = \frac{\Delta a}{a}r \tag{4.59}$$

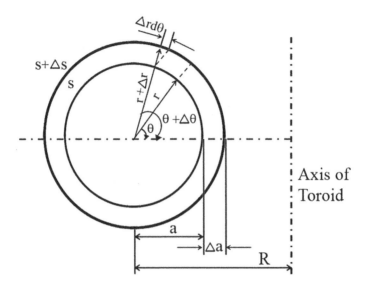

FIGURE 4.13
Virtual model for evaluating the structure required for compensating the net magnetic field.

Substituting Equation (4.59) in Equation (4.58),

$$\int T_{st}\left(\Delta r d\theta\right) = a\frac{\partial E_s}{\partial a} \tag{4.60}$$

The tension or compression in the structural mass is given by:

$$m_a = \frac{\rho_{st}}{\sigma_{st}}\int T_{st} r d\theta = -\frac{\rho_{st}}{\sigma_{st}} a\frac{\partial E_s}{\partial a} \tag{4.61}$$

Using the inductance and Equation (4.61)

$$m_a = q_a\frac{\rho_{st}E_s}{\sigma_{st}} \tag{4.62}$$

where,

$$q_a = \left(\frac{-\beta}{f}\frac{\partial f}{\partial \beta}\right) \tag{4.63}$$

From Equation (4.63) and inductance of the superconductor, the total structural mass required, depending on the radial forces and circumferential forces, is given by:

$$m_{st} = \frac{\rho_{st}E_s}{\sigma_{st}}\left[q_a + q_R\right] \geq \frac{\rho_{st}E_s}{\sigma_{st}} \tag{4.64}$$

$$q_R + q_a = -\left(1 - \frac{\beta}{f}\frac{\partial f}{\partial \beta}\right) + \left(\frac{-\beta}{f}\frac{\partial f}{\partial \beta}\right) \tag{4.65}$$

$$q_R + q_a = \begin{cases} \left|1 + 2\left(\dfrac{\beta}{f}\dfrac{\partial f}{\partial \beta}\right)\right|, \text{ for } \dfrac{\beta}{f}\dfrac{\partial f}{\partial \beta} \leq 0 \\[4mm] 2\dfrac{\beta}{f}\dfrac{\partial f}{\partial \beta}, \text{ for } \dfrac{\beta}{f}\dfrac{\partial f}{\partial \beta} \geq 1 \end{cases}$$

In Equation (4.65), consider that $\dfrac{\beta}{f}\dfrac{\partial f}{\partial \beta} = q_c$

Equation (4.65) can be simplified as:

$$m_{st} = \frac{\rho_{st} E_s}{\sigma_{st}}(1 + 2q_c) \tag{4.66}$$

The magnetic energy stored in the conductor entirely depends on the structural mass of the superconductor.

$$E_m = \frac{\sigma_{st}}{\rho_{st}}(m_t - m_c) \tag{4.67}$$

This is the expression that gives the relation between the 1D stresses (tension and compression) and the structural masses.

$$m_{structure} = m_t + m_c = 2m_c + E_m \frac{\rho_{st}}{\sigma_{st}} \tag{4.68}$$

The ultimate limit in the system is possible when the structural masses experience tension.

$$m_{min} = E_m \frac{\rho_{st}}{\sigma_{st}} \tag{4.69}$$

$$E_m = k_t \frac{\sigma_{st}}{\rho_{st}}(m_{total}) \tag{4.70}$$

where, σ_{st} denotes working stress, ρ_{st} is density of structural mass and m_t and m_c are tension and compression of mechanical structure. k_{st} is the factor that depends on the topology of the SMES.

$$k_{st} = \begin{cases} 0.334 & \text{for infinite thin and long solenoid} \\ >0.334 & \text{for short solenoid} \end{cases}$$

$$k_{st} = \begin{cases} 0.334 & \text{for infinite thin toroid with circular section} \\ <0.334 & \text{for real toroid with circular section} \end{cases}$$

If the solenoid superconducting coil topology is used, hoop stresses are induced in the coil. The maximum hoop stress induced in the solenoid coil is given by:

$$\sigma_{st} = \frac{J(\mu_0 J_e) R_T}{2} \quad \left[\because B_{max} = \mu_0 J_e \right] \tag{4.71}$$

For thin solenoids, the energy stored can be calculated by:

$$E_m = \frac{1}{2\mu_0} B_{max}^2 \pi R_T^2 h \tag{4.72}$$

where, h = height of thin solenoid.

Based on the hoop stresses, the energy stored in the solenoid is given by:

$$E_m = \frac{\sigma}{2} V_s \tag{4.73}$$

Based on the compressive forces, the energy stored in the solenoid is given by:

$$E_m = \frac{\sigma}{3} V_s \tag{4.74}$$

where, V_s is volume of solenoid.

4.3 Cryogenic Refrigeration Systems

Efficient operation of SMES is highly dependent on the energy storage capability of the system. In order to operate the SMES efficiently, a cryogenic refrigeration system is necessary to maintain the temperature of the superconducting coil lower than the critical temperature of the superconducting tapes or wires at higher magnetic fields. The design of the cryogenic system of SMES includes a refrigerator, where the cryogenic coolant is prepared; a vacuum sealed cryostat, where the superconducting coils are cooled and thermally isolated from the ambient, distribution system; and a pair of cryogenic liquefiers. The necessity of the distribution system is to integrate the refrigeration system, cryogenic control valves, vacuum devices, cryostat, liquefiers, high-pressure relief valves and cryogenic reserve pot with the electrical connection system. The commercially available SMES are operated at a temperature of 4.5 K. Forced cooling or bath cooling are used for cooling the SMES, and helium (T_c = 4 K) is employed as a cryogenic coolant to minimize the heat load on the cryostat. Compressors are used in the refrigeration system to provide gaseous helium to the vacuum sealed cold box at an ambient temperature, and the superconducting coil is cooled by the produced liquid helium.

4.3.1 Heat Loads on Cryogenic Refrigeration System

The refrigeration system used for cooling the SMES encounters heat loads from five different sources. The efficiency of the SMES depends on the efficient reduction of the heat loads to ensure the non-dissipative superconducting coil operation to store high capacity. The losses that develop the heat load on the cryogenic refrigeration system include [42–44]:

- Eddy current losses around the superconductors.
- Hysteresis losses due to field penetration in the superconductors.
- Thermal losses at the junction of cold to warm current leads.
- Thermal conduction between the cryostat and the support.
- Thermal radiation from the vacuum vessel.

4.3.1.1 Eddy Current Losses

The superconducting coil consists of composite superconductors that include filaments of superconducting material in the Cu, Al or Ag matrix. The filaments in the superconducting wire are not aligned parallel to each other; however, these are twisted or wound to reduce losses due to eddy currents. These wires are transposed over the length of the mandrel that defines the twist pitch. The eddy losses are proportional to the square of the twist pitch of the superconductor.

$$Q_E = \frac{0.25B^2}{\mu_0 \tau}\left(1+(\tau\omega)^{-2}\right)^{-1} \tag{4.75}$$

where, μ_0 is permeability of tape in free space, B is magnetic flux density, ω is angular frequency and τ is time decay constant of the induced electrical current.

4.3.1.2 Hysteresis Losses

These losses are due to the penetration of the self field in the twisted composite superconductors and are proportional to the rate of change in the magnetic field and twist pitch and are inversely proportional to the radius of the superconductor. Since these losses depend on the dimensions of the superconducting wire, replacing the wire with small wires can reduce losses.

From the bean model, average losses in the HTS tapes per unit length can be estimated as follows,

$$Q_H = 4\mu_0 A_s H_{mag} H_{pk} \left(1 + \left(\frac{I}{I_C}\right)^2\right) \tag{4.76}$$

where, A_s is area of the tape [$A_s = wt$], w and t are the width and half thickness of the tape, H_{mag} is behavior of outer magnetic field and H_{pk} is peak value of the generated magnetic field. I and I_C are the operating and critical currents, respectively.

4.3.1.3 Current Lead Losses

During charging and discharging of the superconducting coil, the PCS becomes warm due to the associated electronic circuits, and the thermal losses occur at the junctions. These losses occur continuously if the superconducting coil is operated without a cold switch. The losses associated with the current leads are a summation of the joule heat and conduction heat. The losses through the leads can be calculated as

$$q_{lead} = I_T \left(\sqrt{L_C (T_{amb} - T_C)}\right) \tag{4.77}$$

$$P_{lead} = I_{lead} q_{lead} \tag{4.78}$$

where, I_T is transport current, L_C is Lorentz constant (2.445×10^{-8} W.Ω/K^2), T_{amb} is ambient temperature (300 K), T_C is cryogenic coolant temperature, I_{lead} is current in the lead and q_{lead} is heat transfer rate through the lead and it depends on the design.

4.3.1.4 Conductive Losses

The cryogenically cooled superconducting coil in the vacuum sealed cryostat is supported by the mechanical structure which is at a higher temperature than the superconducting coil. Hence, conductive losses occur at the junction of the mechanical support (strut) and the cryostat. The heat loads due to conductive losses are given by the total struts used for supporting the cryostat and are directly proportional to the cold mass used in the cryostat under the stresses when the strut remains constant. The magnitude of these losses is very small. The conductive losses from the superconducting coil can be estimated from Fourier's law of heat transfer

$$q_{\text{cond}} = \frac{A}{L} \int\limits_{T_C}^{T_H} k(T)\,dT \qquad (4.79)$$

where, A and L are area and length of the heat transfer surface, $k(T)$ is temperature-dependent thermal conductivity of the superconducting material and T_H and T_L are the high temperature and low temperature sources.

4.3.1.5 Radiation Losses

The heat that is radiated from the vacuum sealed barrier by black body radiation at a given rate is termed as radiation power density. The emitted radiation in the SMES is absorbed by the cold mass; the product of power density and warm area constitutes the total power absorbed. These can be reduced by using the thin intermediate layers of super insulation to shield the radiation losses from the warm surface to the cold mass.

$$Q_R = A_F \sigma \varepsilon \left(T_{\text{Shield}}^4 - T_{SC}^4 \right) \qquad (4.80)$$

where, A_F is external surface area of the superconducting coil, ε is emissivity of the superconducting coil, σ is Stefan–Boltzmann constant (5.67×10^{-8} W/m^{2}K^4) and T_{SC} and T_{shield} are temperatures of the superconducting coil and shield, respectively.

4.3.2 Efficiency of Refrigeration System

Enormous efforts are essential while designing the cryogenic systems and SMES to reduce the losses in the superconducting coils, and therefore the heat load on the cryogenic environment will be minimized from all heat sources. To maximize the efficiency of such a system, the heat loads from the mechanical supports, supply leads and the radiation from the system should be lower. These heat loads can be minimized by installing "thermal shields" that provide intermediate cooling. These heat loads are considered while designing the cryogenic system, because the heat loads generated by SMES during operation are comparatively lower.

The cryogenic refrigeration system utilizes external power in the form of electricity to reduce the heat loads. Hence, the efficiency of the SMES decreases. To evaluate the total efficiency of the SMES, the refrigeration cycle efficiency is essential. The efficiency of the refrigeration cycle is given by,

$$\eta_{\text{cycle}} = \frac{E}{E + \left(P t_{\text{cycle}} \right)} \qquad (4.81)$$

where, E is energy stored, t_{cycle} is duration of cycle and P is power consumption of the cryogenic refrigerator and is 300–1000 times of the cooling power.

The operation of SMES is complex with technical problems such as thermal insulation to minimize the heat loads and the superconducting coil refrigeration. Further, extremely low temperatures (approximately 1.8 K) are required to enhance the current-carrying capability of the superconductor. Superfluid helium II can provide an advantage of enhanced heat transfer for such technical problems [1]. To prevent heat transfer from the ambient to the cryogenic environment, superconducting coils in circular cryostat vessels need to be vacuum sealed. Further, the mechanical stresses induced in the superconducting coil should be transmitted to the low thermal conductance of the cryostat support (strut). The issues related to tensile forces in the superconductors can be reduced by employing the corrugated arrangement of the vessel walls and conductor, thereby providing the movement for the conductor caused by thermal contraction as well as magnetic pressure. Furthermore, for such issues, a cryostat with thin walls containing thermal shields is required, and each shield should be cooled at different temperatures to maintain the extremely low temperature (approximately 1.8 K) to reduce the heat loads.

4.3.3 Cooling Methods

The efficient operation of the superconducting coil depends on the temperature of the cryogenic coolant that is circulated. A cold source is used to impose the cryogenic temperatures designed earlier to the design of the SMES coil. The mode of cooling and the cooling media used for cooling the SMES at different temperatures are given in Table 4.4.

4.3.3.1 Liquid Bath Cooling

Bath cooling system is the simplest cooling technique that employs direct cooling mechanism in which the cryogenic coolant is in direct contact with the superconductor either by filling the coolant in the cryostat or by flowing through the integrated channel of the device. Convection mode of heat transfer is dominant in such cooling systems. In a bath cooling system, the superconducting coil is submerged into the liquid cryogenic coolant and the heat loads are deposited inside the bath. The cryogenic bath can be maintained in thermal equilibrium either by the saturated or subcooled cryogenic coolant. When the temperature of the cryogenic coolant exceeds the critical temperature due to boiling, the vapor coolant is removed from the bath to enhance the heat transfer rate and to reduce the vapor pressure on the cryostat. Excessive heat load and sudden heating of the bath result in the sharp decrement in the magnetic flux, quenching of the superconductor and cryostat vacuum insulation losses.

In bath cooling of the SMES system, two traditional cooling methods are employed, dip cooling and forced cooling (see Figure 4.2), that are used to cool the superconductor below the critical temperature. In this method,

TABLE 4.4

Different Cooling Systems and Cooling Media Used for Cooling SMES

	Cooling system	
Cooling media	Open cycle cooling	Closed cycle cooling
Gas and fluid	The heat exchange takes place due to the phase change of the cryogenic coolant used and is most likely suitable for the helium cooling. This cooling is used when the temperature range is from 1.2 K to 10 K.	The helium gas used for cooling is re-liquefied and re-circulated continuously.
Liquid bath	This cooling method is majorly employed due to ease in construction and maintenance. The cryogenic coolant used determines the temperature range. Cryogenic coolants such as helium is limited to 4.2 K, nitrogen to 77 K, hydrogen to 20 K and neon to 27 K. However, neon is difficult to isolate in electrical environment and is expensive, whereas hydrogen is too dangerous.	This system of cooling consumes high volumes of energy, voluminous compared to the open cycle cooling. However, this system is autonomous.
Conduction	This mode of cooling is suitable for high-temperature superconducting coils that are operated at temperatures higher than 15 K. At lower temperatures, the temperature gradients induced by the superconducting coil due to conduction cooling is problematic for the devices that are operated at 4 K.	

cryogenic coolants such as liquid helium or liquid nitrogen are used. During the charging and discharging of SMES losses the PCS, AC losses (self and external) are generated by the superconducting coil; conductive and radiative losses in the vacuum sealed cryostat leads to heat loads on the cryogenic system. The heat loads encountered by the SMES are stabilized by the cryogenic coolant by gaining the heat to protect the superconducting coil from run-off. The dip cooling system has the potential advantage of thermal stability of the system. However, the electrical stability for voltage proof and AC losses is not possible. Figure 4.14 shows the schematic of the dip cooling of the SMES. Forced cooling is better compared to dip cooling in terms of performance, compensation of AC losses and voltage proof. The stability offered by forced cooling is relatively less than that in the dip cooling system. Further, a complex cryogenic cooling system and continuous circulation of cryogenic coolant is required for both cooling methods to operate SMES for a long time. The volumetric flow rate of the cryogenic coolant depends on the heat load on the cryogenic system and the boil off rate. To overcome such problems for cooling the SMES, utilization of cryocooler is a better alternative than liquid helium cooling due to its safety, efficiency in operation and low pressure liquefaction. In the cryocooler, the superconductor is cooled by the cryogenic circulation loop through conduction mode of heat transfer in the superconducting coil and convection mode of heat transfer in the circulation loop [2].

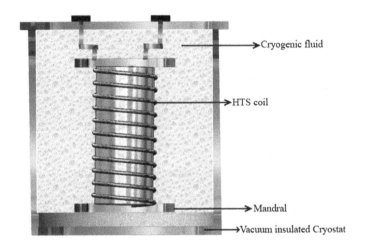

FIGURE 4.14
Liquid bath cooling of SMES.

4.3.3.2 Steps for Solving the Heat Transfer Rate in the Bath Cooling System

The heat transfer from the superconducting coil to the cryogenic cooling system is dependent on the hydraulic characteristic length (cross-sectional area to wetted perimeter), thermophysical properties of the cryogenic coolant, heat efflux to the cryogenic coolant from the superconductor and nature and geometry of the superconducting device. Since the heat transfer in the bath cooling is due to natural convection and boiling, the bulk transport fluid (cryogenic coolant) experiences the heat flux from the surface of the superconductor.

1) The heat deposited in the cryogenic bath is proportional to the cryogenic coolant evaporation rate and can be expressed as

$$\dot{q} = \left(\dot{m} L \right)_{LN_2} \tag{4.82}$$

where, $\dot{q}$ is the heat loads from different sources deposited in the cryogenic bath, $\dot{m}_{LN_2}$ is cryogenic coolant evaporation rate and L_{LN_2} is latent heat of vaporization.

2) Calculate the bulk mean temperature of the cryogenic coolant

$$T_{b,m} = \frac{T_{CC} + T_S}{2} \tag{4.83}$$

where, $T_{b,m}$ is bulk mean temperature of the cryogenic coolant, T_{CC} is temperature of the cryogenic coolant and T_S is surface temperature of the superconducting device.

3) Calculate the volume expansion coefficient of the cryogenic coolant

$$\beta = \frac{1}{T_{b,m}} \tag{4.84}$$

4) Calculate the Grashof number

It is the ratio of the buoyancy forces to the viscous forces that are acting in the flow field. This is the important parameter used to determine the convective heat transfer rate.

$$Gr = \frac{\rho(T)^2 g\beta L_c^2 (T_S - T_{CC})}{\mu(T)^2} \tag{4.85}$$

where, g is gravitational constant, β is volume expansion coefficient, L_C is characteristic length, $\mu(T)$ is temperature-dependent viscosity of the cryogenic coolant and $\rho(T)$ is temperature-dependent density of the cryogenic coolant.

5) Calculate the Prandtl number

It is the ratio of momentum diffusion to thermal diffusion. This is also the parameter used to determine the convective heat transfer rate.

$$Pr = \frac{\mu(T)C(T)}{k(T)} \tag{4.86}$$

where, $C(T)$ is temperature-dependent specific heat of cryogenic coolant and $k(T)$ is temperature-dependent thermal conductivity of cryogenic coolant.

6) Calculate the Nusselt number

$$Nu = \frac{hL_C}{K(T)} = A(Gr\,Pr)^m \tag{4.87}$$

where, h is convective heat transfer coefficient and A and m are constants depending on different parameters of the superconducting devices.

7) The heat transfer rate due to convection can be calculated as follows using Equation (4.87)

$$\dot{q} = h(T_S - T_{CC}) \tag{4.88}$$

Otherwise, equating, Equation (4.82) and Equation (4.88)

$$h = \frac{(\dot{m}L)_{LN_2}}{(T_S - T_{CC})} \tag{4.89}$$

Substituting Equation (4.89) in Equation (4.88), the heat transfer rate can be calculated.

When the surface of the superconducting coil increases above the boiling point of the cryogenic coolant, the formation of bubbles starts along with the boiling of the cryogenic coolant. This is due to the rapid increase in the convective heat transfer rate of the cryogenic coolant. The bubble formation in the bath reduces as the heat transfer rate decreases drastically. This is an important phase in the practical conditions because boiling of the coolant can lead to quenching of the superconductor.

8) The heat transfer rate during boiling can be calculated as follows

$$\dot{q}_{boiling} = h_b \left(T_S - T_{sat} \right) \tag{4.90}$$

where, h_b is convective heat transfer rate during boiling and T_{sat} is the saturation temperature of the cryogenic coolant. For instance, the saturation temperature of the LN_2 at 1 atm is 77 K.

4.3.3.3 Conduction Cooling

In conduction cooling, the cold cryogenic source is directly connected to the superconducting coil using external thermal links. The mode of heat transfer for cooling the superconducting coil is pure conduction. This is the robust method of cooling the superconducting coil and has better stability than bath cooling. Figure 4.15 shows the schematic of the conduction-cooled cooling method for cooling the superconducting coil [high-temperature superconductivity-2]. The operating temperature depends on the type of superconductor used in the design of the superconducting coil. For instance, if the superconducting coil is designed using high-temperature superconductors such as BSCCO or YBCO, then the operating temperature of the cryogenic coolant can be maintained above 20 K and the SMES can be operated around 50–60 K. In conduction cooling method, a pulse-tube or Gifford–McMahon (GM) cryocooler is utilized as a heat sink to cool the superconducting coil below the critical temperature of the superconductor [42, 45]. The accommodation of a cryocooler next to a vacuum sealed cryostat that consists of superconducting coil is not advisable, because the cooling performance will be degraded. Further, the conduction heat transfer rate is poor due to the complexity in thermal contact between the superconducting coil and the cryocooler, which is considered as the critical factor while designing the conduction cooling system. Furthermore, the dimensions of superconducting coil cannot be altered for attaining higher capacities because there is an increase in the magnitude of the magnetic field, resulting in an extensive cooldown time that is required for cooling the superconducting coil.

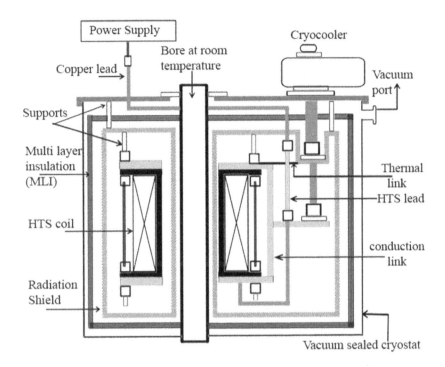

FIGURE 4.15
Direct cooling of SMES.

The transient heat conduction equation applicable for the superconducting coil at the cryogenic temperature can be calculated using the energy conservation equation:

$$\nabla \cdot \left(-k(T)\nabla T\right) + \dot{q}_g = \rho C \frac{\partial T}{\partial t} \qquad (4.91)$$

where, $k(T)$ is temperature-dependent thermal conductivity of the superconducting coil, $\dot{q}_g$ is heat generated in the superconducting coil due to AC losses and ρ and C are density and specific heat of the superconducting coil, respectively. The first term in the left-hand side represents the conduction term, the second term indicates the heat generation in the superconducting coil and the term on the right-hand side indicates the thermal inertia term.

If the heat transfer is considered in one dimension, the above equation can be represented as:

$$\frac{\partial^2 T}{\partial x^2} + \frac{1}{k}\dot{q}_g = \frac{\rho C}{k}\frac{\partial T}{\partial t} \qquad (4.92)$$

From the Fourier's law of heat conduction, the heat transfer rate due to conduction can be written as:

$$\dot{Q}_{cond} = \frac{A_{cond}}{L_{cond}} \int_{T_{cold}}^{T_{hot}} k(T) dT \tag{4.93}$$

where, A_{cond} and L_{cond} are cross-sectional area and the length of the superconductor used in the SMES coil, respectively.

4.3.4 Thermal Issues Related to Cooling of SMES

In this section, the thermal issues related to heat loads encountered by the SMES and the mathematical expressions related to the cooling of such issues are discussed. This section provides a clear idea of designing the SMES efficiently to counter the heat loads.

i. **Superconducting coil**

The superconducting coils are cryogenically cooled using cryogenic coolants such as helium (liquid and gaseous) and nitrogen (liquid). These coils at the cryogenic environment encounter thermal issues while charging and discharging the electrical current from and to the grid. They are modes for cooling the coil and helium gas cooled leads. Major sources of heat loads on the cryogenic coolants are *joules heating of the superconducting coil* during the electric current flow and *axial conduction cooling at the junctions of the leads* between the surrounding and the cryogenic environment. There are two modes of cooling a superconducting coil [46]. They are *perfect cool down* and *dunk mode*. Let us consider the perfect cool down mode. In this mode, the superconducting coil is cooled by liquid nitrogen (LN$_2$), using a series of infinitesimal energy exchangers.

a) **Perfect cool down mode**

For a perfect cool down, the heat rejected by the superconducting coil should be equal to the heat gained by liquid nitrogen:

$$dq_{SC} = dq_{LN_2} \tag{4.94}$$

$$m_{SC} C_{SC}(T) dT = dm_{LN_2}[h_{LN_2}(T) - h_{LN_2}(77\,K, \text{liquid})] \tag{4.95}$$

where, m_{SC} is mass of the superconducting coil, $C_{SC}(T)$ is temperature-dependent specific heat of superconducting coil, m_{LN_2} is mass of LN$_2$ circulated for cooling the coil, $h_{LN_2}(T)$ is temperature-dependent specific enthalpy of LN$_2$ at the temperature T and $h_{LN_2}(77\,K, \text{liquid})$ is specific enthalpy of wet saturated LN$_2$ at 77 K.

The mathematical expression for the mass of LN_2 circulated per mass of the superconducting coil is given by:

$$\frac{m_{LN_2}}{m_{SC}} = \int\limits_{77K}^{T} \frac{C_{SC}(T) \times dT}{h_{LN_2}(T) - h_{LN_2}(77K, \text{liquid})} \tag{4.96}$$

b) **Dunk mode**

If the latent heat of vaporization of LN_2 is used to cool the superconducting coil, such a cooling mode is called the dunk mode of cooling.

$$m_{SC} \int\limits_{77K}^{T} C_{SC}(T)dT = m_{LN_2} L_{LN_2} \tag{4.97}$$

$$\frac{m_{LN_2}}{m_{SC}} = \frac{h_{SC}(T) - h_{SC}(77K)}{L_{LN_2}} \quad \left[\because h = C\,dT \right] \tag{4.98}$$

where h_{SC} is the specific enthalpy of the superconducting coil.

ii. **Gas cooled leads**

In leads, the heat loss occurs due to conduction at the junctions, considering that the heat conduction through lead is along the Z direction. Figure 4.16 shows the heat balance of a gas cooled lead in differential volume.

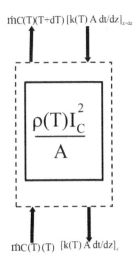

FIGURE 4.16
Heat balance of a gas cooled lead in differential volume.

From the generalized heat conduction equation,

$$Q_{out} - Q_{in} + \frac{\dot{Q}_{gen}}{k} = \frac{\rho V C}{k} \frac{dT}{dt} \tag{4.99}$$

where, Q_{out} is total power flowing out, Q_{in} is total power flowing in, $V = A\Delta Z$ is volume of the lead, k is thermal conductivity of lead and $\dot{Q}_{gen} = \frac{\rho I^2}{V}$ is heat generated per unit volume.

The total power flowing out from the differential volume of the power lead is:

$$Q_{out} = \left[k(T)A\left(\frac{dT}{dz}\right) \right]_z + \dot{m}C(T)(T+dT) \tag{4.100}$$

The total power flowing in from the differential volume of the power lead is:

$$Q_{in} = \left[k(T)A\left(\frac{dT}{dz}\right) \right]_{z+\Delta z} + \dot{m}C(T)(T) + \left(\frac{\rho(T)I_c^2}{A\,\Delta z} \right) \tag{4.101}$$

Under a steady-state condition, the total power flowing out should be equal to the total power flowing in through the differential volume:

$$Q_{in} = Q_{out} \tag{4.102}$$

$$\left[k(T)A\left(\frac{dT}{dz}\right) \right]_{z+\Delta z} + \dot{m}C(T)(T) + \left(\frac{\rho(T)I_c^2}{A\,\Delta z} \right)$$
$$- \left[k(T)A\left(\frac{dT}{dz}\right) \right]_z + \dot{m}C(T)(T+dT) = 0 \tag{4.103}$$

Solving the above equation:

$$\frac{d}{dz}\left[k(T)A\left(\frac{dT}{dz}\right) \right] - \dot{m}C(T)\left(\frac{dT}{dz}\right) + \left(\frac{\rho(T)I_c^2}{A} \right) = 0 \tag{4.104}$$

The heat transfer is gained by the LN_2 bath through conduction when the superconducting coil is operated under high electric current flow. In Equation (4.104), the first term corresponds to the conduction along the length of the lead, and the losses are considerably lower than other heat loads. Simplifying Equation (4.104) by neglecting the heat conduction equation at the junction of current lead and LN$_2$ bath is as follows:

$$-\dot{m}C(T_0)\left(\frac{dT}{dz}\right)_{z=0} + \left(\frac{\rho_0 I_c^2}{A}\right) = 0 \tag{4.105}$$

Solving the temperature gradient, Equation (4.105) can be written as:

$$\left(\frac{dT}{dz}\right)_{z=0} = \left(\frac{\rho_0 I_c^2}{\dot{m}C(T_0)A}\right) \tag{4.106}$$

The heat conducted at the junction of the lead and LN$_2$ bath is:

$$Q = k_0 A\left(\frac{dT}{dz}\right)_{z=0} \tag{4.107}$$

Substituting Equation (4.106) in Equation (4.107),

$$Q = \left(\frac{k_0 \rho_0 I_c^2}{\dot{m}C(T_0)}\right) \tag{4.108}$$

where, k_0 is thermal conductivity and ρ_0 is resistivity of the lead at $z = 0$.

The heat conducted through the lead should be gained by the cryogenic coolant. The amount of heat transfer through the cryogenic coolant (LN$_2$) due to the latent heat of vaporization can be calculated by equating the heat conducted at the junction of the lead and the rate of energy taken to boil off the cryogenic coolant.

$$Q = \dot{m}L_{LN_2} \tag{4.109}$$

Equating Equation (4.108) and Equation (4.109) to get the boil-off rate of LN$_2$,

$$\dot{m} = I_c\sqrt{\frac{k_0\rho_0}{C(T_0)\times h_L}} \tag{4.110}$$

Substituting Equation (4.110) in Equation (4.108):

$$Q = \frac{k_0\rho_0 I_c^2}{I_c\sqrt{\dfrac{k_0\rho_0}{C(T_0)\times h_L}}\,C(T_0)} \tag{4.111}$$

Simplifying Equation (4.111):

$$\frac{Q}{I_c} = \sqrt{\frac{h_L k_o \rho_o}{C(T_0)}} \tag{4.112}$$

This is the expression for the ratio of the total heat gained by the LN_2 bath by the conduction mode of heat transfer to the designed transport current through the superconducting coil.

From Equation (4.105), assuming that $C(T) \approx C$ and integrating with the limits of T_0 at $z = 0$ and T_l at $z = l$:

$$\int_{T_0}^{T_l} \frac{dT}{\rho(T)} = \int_0^l \frac{I_c^2 dz}{A \dot{m} C} = \frac{I_c^2 l}{A \dot{m} C} \tag{4.113}$$

Substituting Equation (4.110) in Equation (4.113):

$$\int_{T_0}^{T_l} \frac{dT}{\rho(T)} = \left(\frac{I_T l}{A}\right)_{ot} \sqrt{\frac{h_L}{\rho C k}} \tag{4.114}$$

Since one end of the lead is exposed to the ambient environment and the other end to the helium environment, the operating temperature limits can be considered as $T_0 = 6$ K and $T_l = 273$ K to obtain the appropriate design ratio.

From Equation (4.104), if heat generation in the lead is considered to be negligible:

$$\frac{d}{dz}\left[k(T)A\left(\frac{dT}{dz}\right)\right] - \dot{m}C(T)\left(\frac{dT}{dz}\right) = 0 \tag{4.115}$$

Assume that $\dfrac{dT}{dz} = \theta(z)$:

$$k(T)A\left(\frac{d\theta}{dz}\right) - \dot{m}C(T)\theta = 0 \tag{4.116}$$

Solving the above equation:

$$\frac{dT}{dz} = \theta = \theta_o \exp\left(\frac{\dot{m}C_0 z}{Ak}\right) \tag{4.117}$$

Integrating the above equation:

$$T(z) = A_0 \exp\left(\frac{\dot{m}C_0 z}{Ak}\right) + A_1 \tag{4.118}$$

Applying the boundary conditions:

$$T(z=0) = T_o = A_o + A_1 \text{ and } \left(\frac{dT}{dz}\right)_{z=0} = \frac{\dot{m}_0 h_l}{Ak_0}$$

Substituting the boundary conditions in Equation (4.118):

$$T(z) = T_O + \frac{h_l k}{Ck_o}\left[\exp\left(\frac{\dot{m}_0 Cz}{Ak}\right) - 1\right] \tag{4.119}$$

This is the mathematical expression for temperature distribution along the length of the lead when the boil-off of LN$_2$ occurs.

If the boundary condition at the base of the lead is applied, i.e., $z = 1$, then $T(z) = T_l$.

Now, Equation (4.119) can be rewritten as:

$$T_l = T_O + \frac{h_l k}{Ck_o}\left[\exp\left(\frac{\dot{m}_0 Cl}{Ak}\right) - 1\right] \tag{4.120}$$

$$\frac{\dot{m}_0 Cl}{Ak} = \ln\frac{Ck_o(T_l - T_o)}{h_l k} + 1 \tag{4.121}$$

Solving the above equation:

$$\dot{m}_0 = \left(\frac{Ak}{Cl}\right)\left(\ln\frac{Ck_o(T_l - T_o)}{h_l k} + 1\right) \tag{4.122}$$

Form the latent heat transfer:

$$Q = \dot{m}_0 L_{LN2} \tag{4.123}$$

Substituting Equation (4.122) in Equation (4.123):

$$Q = L_{LN2}\left(\left(\frac{Ak}{Cl}\right)\left(\ln\frac{Ck_o(T_l - T_o)}{h_l k} + 1\right)\right) \tag{4.124}$$

This expression is the boil off rate or the heat input at the base of the lead.

If heat generation is considered in addition to the heat input at the base of the lead, then the heat transfer rate with internal heat generation is given by:

$$Q_0 = k_o \times \frac{\rho_0 \times I_c^2}{C(T_0)}\frac{1}{I_t}\sqrt{\frac{Ch_L}{k_o\rho_o}} \tag{4.125}$$

Dividing Equation (4.124) by Equation (4.125):

$$\frac{Q}{Q_0} = \frac{kh_l}{\left(1.2\times 10^{11}\right)k_0 C_0 \rho_0}\left(\ln\frac{Ck_0\left(T_l - T_0\right)}{h_l k} + 1\right) \tag{4.126}$$

iii. Optimum design of lead

The time-dependent power equation for the optimum lead can be written as:

$$AC_{cu}(T)\frac{dT}{dt} = \frac{d}{dz}\left[\frac{Ak(T)dT}{dz}\right] - \dot{m}C(T)\frac{dT}{dz} + \rho_{cu}(T)\frac{I_c^2}{A} \tag{4.127}$$

where, C_{cu} and ρ_{cu} are heat capacity and resistivity of copper lead, respectively.

Neglecting the conduction through the copper lead and cooling using the gaseous helium, Equation (4.127) can be written as,

$$AC_{cu}(T)\frac{dT}{dt} = \rho_{cu}(T)\frac{I_c^2}{A} \tag{4.128}$$

Simplifying the above equation for time-dependent temperature:

$$\frac{dT}{dt} = \frac{\rho_{cu}(T)}{C_{cu}(T)}\frac{\varsigma_o^2}{l^2} \tag{4.129}$$

where, $\varsigma_o = \left(\dfrac{I_t l}{A}\right)_{Optimum} = 2.5\times 10^7\ \text{A/m}$

The flow stoppage meltdown is a phenomenon where the current lead gets melted at the top when no coolant is flowing across it and the lead is carrying the operating current. So assuming $C_{cu}(T) = C_0$ (a constant) and $\rho_{cu}(T) = \rho_0 + bT$ (where b is another constant). Hence, Equation (4.129) can be rewritten as:

$$\frac{dT}{dt} = \rho_0\frac{\varsigma_0^2}{l^2 C_0} + \frac{b\varsigma_0^2 T}{C_0 l^2} \tag{4.130}$$

$$\frac{dT}{dt} = \frac{\rho_0}{b\tau_l} + \frac{T}{\tau_l} \tag{4.131}$$

where $\tau_l = \dfrac{C_0 l^2}{b\varsigma_0^2}$, is called the thermal time constant of the lead.

iv. *Gas Cooled support rods*

Structured support rods inside the cryostat represent a conductive heat load on the cryogenic environment. Like the cooling of leads, the cooling of support rods also presents a challenge. The temperature variation along the thickness is T_0 at the bottom end and T_l at the top of the rod. Assume the rod has a cross-sectional area A and the distance between the bottom end and the top end is l. Assume its thermal conductivity is temperature dependent and is given by k. So the conduction heat input through a gas cooled support rod is given by:

$$Q_0 = \frac{h_l k}{C_0}\left(\frac{A}{l}\right)\ln\left(\frac{C_0 k_0 (T_l - T_0)}{h_l k} + 1\right) \tag{4.132}$$

For a non-cooled support rod of cross section A and length l, with a linear temperature variation of (T_0 at z = 0 and T_l at z = l) and constant thermal conductivity of k, the conduction heat input would be:

$$Q_g = Ak\frac{T_l - T_0}{l} \tag{4.133}$$

Therefore,

$$\frac{Q_g}{Q_0} = \frac{C_0 k_0 (T_l - T_0)}{h_l \ln\left(\frac{C_0 k_0 (T_l - T_0)}{h_l k} + 1\right)} \tag{4.134}$$

v. **Residual gas heat transfer into the cryostat**

The residual gas heat input is majorly due to *high pressure* and *low pressure limit*. It is necessary to analyze the amount of residual gases formed in the cryostat due to the heat loads. These gases are formed due to the boiling of the cryogenic coolant. They impart pressure on the cryostat, thereby leading to the failure of vacuum insulation.

a) **High-pressure limit gas residuals**

When the pressure of the gas is sufficiently high, the mean free path of $\lambda_g \ll d$, where d is the typical distance separating the two surfaces at different temperatures in a cryostat; under these conditions the thermal conductivity of a gas k_g, according to kinetic theory is proportional to the mean velocity of the molecule which further depends upon $\sqrt{T}$. In this case, $\lambda_g \ll d$ and k_g are independent of gas pressure, p_g. According to kinetic theory, $\lambda_g \propto \dfrac{T}{P_g}$.

for instance, if liquid nitrogen at T = 300 K and $p_g = 1\text{atm}$, $\lambda_g \approx$ 0.1 μm or the condition of high pressure limit is satisfied. At vacuum, where $p_g \approx 10^{-4}$ torr, λ_g comes out to be 1 m due to which $\lambda_g \ll d$ is violated. So a low pressure limit is required to understand the case.

b)   **Low-pressure limit gas residuals**

In a vacuum condition, k_g becomes proportional to p_g. Hence for a parallel plate configuration having one cold plate at $T_{C,l}$[K] and another warm plate at $T_{W,m}$[K], heat flux from the warm plate to the cold plate by a residual gas at pressure p_g (approximately 10^{-4}torr) is given by:

The rate of heat leakage from residual gases can be estimated as:

$$q_{rg} = c_{air}\beta PS\left(T_{W,m} - T_{C,l}\right) \tag{4.135}$$

where c_{air} is constant of air which is 1.2, β is accommodation coefficient, having a range of 0.59 [42] and P is pressure in the vacuum sealed cryostat.

4.4 Power Conditioning System (PCS)

The power conditioning system is necessary to integrate the SMES with the grid, because for charging and discharging converters are required to convert the AC to DC and from DC to AC, respectively. When such converters are used, harmonics are introduced in the SMES that result in power consumption. Hence, the selection of the PCS is also vital while designing the SMES. The power conditioning system used in the SMES is shown in Figure 4.17.

4.4.1 Power Converters of PCS in the Three-Phase AC Grid

In PCS, the necessity of an inverter or a rectifier in the electronic power circuit is to condition the electrical current as a required output. For instance, the direct current (DC) output from the superconducting coil is converted to alternating current (AC), and vice versa, because the power grids are operated in AC. For conversion of the output current, three topologies are possible [47]:

- **Thyristor bridge** – The active power exchanged cannot be controlled independently and is linked with the reactive power. This converter results in losses in the superconducting coil due to AC losses and harmonics.

$$Q = \frac{4.25}{\pi} U_{rms} I_{SC} \sin\beta \tag{4.136}$$

$$P = \frac{4.25}{\pi} U_{rms} I_{SC} \cos\beta \tag{4.137}$$

where, U_{rms} is RMS voltage amplitude between three phases, I_{SC} is current in the superconducting coil and β is thyristor firing angle.

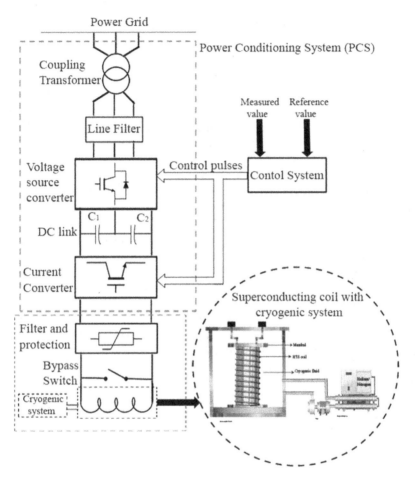

FIGURE 4.17
Power conditioning system in SMES.

- **Voltage source converter** – This is connected in series with a chopper that utilizes insulated gate bipolar transistor (IGBT). A DC bus link is provided between the converters that consist of capacitors. The reactive and active powers are controlled independently using this topology. Further, low AC distortions are observed. This converter results in AC losses due to voltage ripples across the lead of the superconducting coil. If the bus bar is available in the electrical circuit, only the chopper is required in the system.

- **Current source converter** – The reactive and active powers are controlled independently. The SMES is connected to the DC side directly without a chopper in this topology. The AC losses in the superconducting coil are comparatively less in this topology due to fewer harmonics.

4.4.2 Mechanism of Charging and Discharging of SMES

In SMES coil, the charging and discharging are different compared to other storage technologies. In the charging mode, electrical current is conducted by the conversion of alternating current (AC) from the supply or power network to the direct current (DC). Hence, the superconducting coil is exited, and, at this instance, the PCS is available in a rectifier mode. Due to the zero electrical resistance of the superconductor, the electrical energy is directly stored in the superconducting coil without any conversion. In other words, the electrical current flows through the superconducting coil at any state of charge. When the superconducting coil is in charging mode, a positive voltage is produced by the power conversion system which results in the increase in current. When the DC is converted into the AC, the energy stored in the superconducting coil is discharged to the electrical power network. In other words, during the discharge mode, the PCS adjusts the electronics such that a load is applied across the ends of the superconducting coil. Hence, discharge takes place from the superconducting coil due to a negative voltage across the superconducting coil. Hence, due to this characteristic, SMES is considered to be an ideal storage system. The power from the coil can be determined by the product of persistent current and applied voltage across the superconducting coil. For large-scale application of SMES, the issues related to induced electromagnetic fields are very severe because superconducting coils can store higher energy when it is exited with higher electrical current and leads to higher magnetic fields. Hence, the SMES should be designed such that the superconducting coil current and the voltage should be within the allowable margins and safety (Figure 4.18).

4.5 Future Applications of SMES

SMES can be installed in different locations in future grids, depending on the application. These can be an alternative to protect the power systems with uninterrupted power supply during power outage, voltage sag, bulk energy storage, voltage stabilization, load sensitivity etc. showing the future power grid at which SMES can be installed for uninterrupted power [48].

4.5.1 Power Quality

The power quality and uninterrupted power supply (UPS) are crucial in situations where abrupt power fluctuations are possible. For instance, an industrial customer requires a source for uninterrupted power supply from the utility that can compensate the demand utilizing the power reserves until the utility restores the power. In such cases, a system that discharges

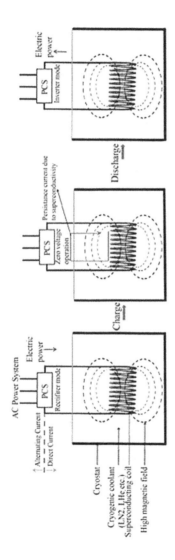

FIGURE 4.18
Mechanism of charging and discharging of SMES.

electrical power to avoid these interruptions in power supply is required and is integrated with electrical power grids. SMES is one such technology that plays a vital role of power back-up and power quality. Further, the disturbances caused due to voltage surge, lightning and power trips at the substations can also be avoided with SMES installations.

4.5.2 System Stability and Frequency Regulation

In long length power applications, power systems encounter unexpected instabilities due to changes in various conditions during power delivery and cause high economic damage. The increase in power demand the conventional power grids integrating with the renewable energy sources such as wind, solar energy etc. However, these sources are located very far from the utility. In this case, if a large-scale plant of wind or solar is considered, the power generated is discontinuous. In such systems, high discharge rate with stability is not possible without storage systems.

SMES technology can provide system stability and the frequency regulation due to its high reliability. Many researches proposed direct installation of SMES coupling, with wind and solar that can increase the efficiency of the plant due to voltage stability. This is possible by charging the superconducting coil during extreme power generation and discharging it during the voltage drop in the grid. Further, the controlled reactive power is advantageous in damping the frequency regulations.

4.5.3 Load Leveling

The variations in electrical power demand are random and predictable. For instance, the power demand of residential and commercial industries, educational institutions etc., will be higher during the day compared to that of the night. To deliver such huge power at high demand requires efficient operation of the power plants with maximum possible power output. A solution for such a problem is to install a power storage system that can charge during the night and discharge the power to the power grid during the peak demand. This is possible by installing an SMES that results in the trade-off between the cost of peak and off-peak demand of power. Further, the installation cost of new power generation systems can be reduced with diurnal storage.

4.5.4 Bulk Energy Storage

The advantage of the SMES over conventional storage system is a series of superconducting coils which are force balanced and can store a huge amount of energy that can be utilized for peak loads at a time of reduced generation capacities. However, the installation cost incurred will be higher than the current storage cost of technologies (Figure 4.19).

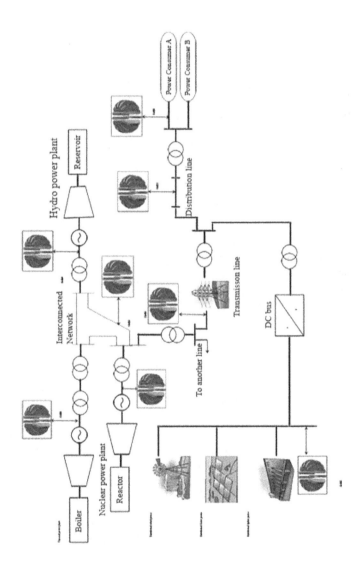

FIGURE 4.19
Installation of SMES in future smart power grid applications.

References

1. A. G. Ter-Gazarian, *Energy Storage for Power Systems*, 2nd edition. London, UK, 2011.
2. P. Seidel, *Applied Superconductivity, Handbook on Devices and Applications*. Weinheim, Germany, 2015.
3. S. Nomura *et al.*, "Experimental results of a 7-T force-balanced helical coil for large-scale SMES," *IEEE Trans. Appl. Supercond.*, vol. 18, no. 2, pp. 701–704, 2008.
4. T. Mito *et al.*, "Development of 1 MJ conduction-cooled LTS pulse coil for UPS-SMES," *IEEE Trans. Appl. Supercond.*, vol. 17, no. 2, pp. 1973–1976, 2007.
5. H. Hayashi *et al.*, "Test results of power system control by experimental SMES," *IEEE Trans. Appl. Supercond.*, vol. 16, no. 2, pp. 598–601, 2006.
6. S. Nagaya *et al.*, "Development and performance results of 5 MVA SMES for bridging instantaneous voltage dips," *IEEE Trans. Appl. Supercond.*, vol. 14, no. 2, pp. 699–704, 2004.
7. S. Nagaya *et al.*, "Field test results of the 5 MVA SMES system for bridging instantaneous voltage dips," *IEEE Trans. Appl. Supercond.*, vol. 16, no. 2, pp. 632–635, 2006.
8. A. Morandi *et al.*, "Design, manufacturing and preliminary tests of a conduction cooled 200 kJ Nb-Ti μSMES," *IEEE Trans. Appl. Supercond.*, vol. 18, no. 2, pp. 697–700, 2016.
9. L. Ottonello *et al.*, "The largest Italian SMES," *IEEE Trans. Appl. Supercond.*, vol. 16, no. 2, pp. 602–607, 2006.
10. X. Jiang, X. Zhu, Z. Cheng, X. Ren, and Y. He, "A 150 kVA/0.3 MJ SMES Voltage Sag Compensation System," *IEEE Trans. Appl. Supercond.*, vol. 15, no. 2, pp. 1903–1906, 2005.
11. Q. Wang *et al.*, "Development of large scale superconducting magnet with very small stray magnetic field for 2 MJ SMES," *IEEE Trans. Appl. Supercond.*, vol. 20, no. 3, pp. 1352–1355, 2010.
12. X. Zhou, X. Y. Chen, and J. X. Jin, "Development of SMES technology and its applications in power grid," *2011 Int. Conf. Appl. Supercond. Electromagn. Devices, ASEMD 2011*, pp. 260–269, 2011.
13. J. D. Rogers *et al.*, "30-MJ superconducting magnetic energy storage system for electric utility transmission stabilization," *IEEE Trans. Magn.*, vol. MAG-15, no. 1, pp. 820–823, Sydney, Australia, 14 Dec 2011, 1979.
14. C. A. Luongo, T. Baldwin, P. Ribeiro, S. Member, and C. M. Weber, "A 100 MJ SMES demonstration at FSU-CAPS," *IEEE Trans. Appl. Supercond.*, vol. 13, no. 2, pp. 1800–1805, 2003.
15. H. J. Kim *et al.*, "3 MJ/750 kVA SMES system for improving power quality," *IEEE Trans. Appl. Supercond.*, vol. 16, no. 2, pp. 574–577, 2006.
16. K. Shikimachi *et al.*, "Development of MVA class HTS SMES system for bridging instantaneous voltage dips," *IEEE Trans. Appl. Supercond.*, vol. 15, no. 2, pp. 1931–1934, 2005.
17. P. Tixador *et al.*, "First tests of a 800 kJ HTS SMES," *IEEE Trans. Appl. Supercond.*, vol. 18, no. 2, pp. 774–778, 2008.
18. K. Koyanagi *et al.*, "Design of a high energy-density SMES coil with Bi-2212 cables," *IEEE Trans. Appl. Supercond.*, vol. 16, no. 2, pp. 586–589, 2006.

19. Q. Wang, S. Song, Y. Lei, and Y. Dai, "Design and fabrication of a conduction-cooled high temperature superconducting magnet for 10 kJ superconducting magnetic energy storage system," *IEEE Trans. Appiled Supercond.*, vol. 16, no. 2, pp. 570–573, 2006.

20. Q. Wang et al., "A 30 kJ Bi2223 high temperature superconducting magnet for SMES with solid-nitrogen protection," *IEEE Trans. Appl. Supercond.*, vol. 18, no. 2, pp. 754–757, 2008.

21. J. Shi et al., "Development of a conduction-cooled HTS SMES," *IEEE Trans. Appl. Supercond.*, vol. 17, no. 3, pp. 3846–3851, 2007.

22. S. Dai et al., "Design of a 1 MJ/0.5 MVA HTS magnet for SMES," *IEEE Trans. Appl. Supercond.*, vol. 17, no. 2, pp. 1977–1980, 2007.

23. W. S. Kim et al., "Design of HTS magnets for a 600 kJ SMES," *IEEE Trans. Appl. Supercond.*, vol. 16, no. 2, pp. 620–623, 2006.

24. H. K. Yeom et al., "An experimental study of the conduction cooling system for the 600 kJ HTS SMES," *IEEE Trans. Applied Supercond.*, vol. 18, no. 2, pp. 741–744, 2008.

25. J. H. Choi, H. G. Cheon, J. W. Choi, H. J. Kim, K. C. Seong, and S. H. Kim, "A study on insulation characteristics according to cooling methods of the HTS SMES," *Phys. C Supercond. Appl.*, vol. 470, no. 3, pp. 1703–1706, 2010.

26. R. Kreutz et al., "Design of a 150 kJ high- T c SMES (HSMES) for a 20 kVA uninterruptible power supply system," *IEEE Trans. Appl. Supercond.*, vol. 13, no. 2, pp. 1860–1862, 2003.

27. J. Kozak, S. Kozak, T. Janowski, and M. Majka, "Design and performance results of first polish SMES," *IEEE Trans. Appl. Supercond.*, vol. 19, no. 3, pp. 1981–1984, 2009.

28. J. Kozak, M. Majka, S. Kozak, and T. Janowski, "Performance of SMES system with HTS magnet," *IEEE Trans. Appl. Supercond.*, vol. 20, no. 3, pp. 1348–1351, 2010.

29. C. J. Hawley and S. A. Gower, "Design and preliminary results of a prototype HTS SMES device," *IEEE Trans. Appl. Supercond.*, vol. 15, no. 2 PART II, pp. 1899–1902, 2005.

30. H. Ueda, A. Ishiyama, K. Shikimachi, N. Hirano, and S. Nagaya, "Stability and protection of coils wound with YBCO bundle conductor," *IEEE Trans. Appl. Supercond.*, vol. 20, no. 3, pp. 1320–1323, 2010.

31. W. Yuan and M. Zhang, "Superconducting magnetic energy storage (SMES) systems," in *Handbook of Clean Energy Systems*, Editor: Jinyue, West Sussex, UK, 2015, pp. 1–16.

32. V. S. Vulusala and S. Madichetty, "Application of superconducting magnetic energy storage in electrical power and energy systems: A review," *Int. J. ENERGY Res.*, vol. 42, no. 2, pp. 358–368, 2017.

33. Y. Horiuchi, Y. Yamasaki, T. Ezaki, and T. Imayoshi, "Solenoid type shielding coil systems for a small scale SMES," *IEEE Trans. Appl. Supercond.*, vol. 18, no. 2, pp. 709–712, 2008.

34. T. Ezaki, Y. Horiuchi, T. Hatanaka, M. Joko, S. Setoguchi, and H. Hayashi, "Toroidal-type shielding coil systems for the power system control SMES," *IEEE Trans. Appl. Supercond.*, vol. 14, no. 2, pp. 746–749, 2004.

35. B. Vincent, P. Tixador, T. Lecrevisse, J. M. Rey, X. Chaud, and Y. Miyoshi, "HTS magnets: Opportunities and issues for SMES," *IEEE Trans. Appl. Supercond.*, vol. 23, no. 3, p. 5700805, 2013.

36. O. Vincent-Viry, A. Mailfert, and D. Trassart, "New SMES coil configurations," *IEEE Trans. Applied Supercond.*, vol. 11, no. I, pp. 1916–1919, 2001.

37. D. Lieurance, F. Kimball, C. Rix, and C. Luongo, "Design and cost studies for small scale superconducting magnetic energy storage (SMES) systems," *IEEE Trans. Appl. Supercond.*, vol. 5, no. 2, pp. 350–353, 1995.

38. Y. Oga, S. Noguchi, and M. Tsuda, "Comparison of optimal configuration of SMES magnet wound with MgB 2 and YBCO conductors," *IEEE Trans. Appl. Supercond.*, vol. 23, no. 3, pp. 0–3, 2013.

39. T. Shintomi, "Applications of High-Tc superconductors to superconducting magnetic energy storage (SMES)," in *High Temperature Superconductivity 2*, Editor: Anant V. Narlikar, Springer-Verlag Berlin Heidelberg, New York, 2004, pp. 213–222.

40. M. Lebioda, J. Rymaszewski, and E. Korzeniewska, "Simulation of thermal processes in superconducting pancake coils cooled by GM cryocooler," *J. Phys. Conf. Ser.*, vol. 494, pp. 1–9, 2014.

41. Y. M. Eyssa and R. W. Boom, "Considerations of a large force balanced magnetic energy storage system," *IEEE Trans. Magn.*, vol. 17, no. 1, pp. 460–462, 1981.

42. X. Song, J. Holbøll, Q. Wang, Y. Dai, N. Mijatovic, and J. Wang, "A conduction-cooled superconducting magnet system - design, fabrication and thermal tests," in *Proceedings of the Nordic Insulation Symposium*, Copenhagen, Denmark 15–17. June 2017, no. 24, pp. 147–151.

43. S. M. Schoenung, R. L. Bieri, and T. C. Bickel, "Superconducting magnetic energy storage (SMES) using high temperature superconductors (HTS) in three geometries," in *Advances in Cryogenic Engineering. Volume 39, part A*, Editor: Peter Kittel, vol. 39, Springer, Boston, Massachusetts, 1993, pp. 829–836.

44. G. Wu *et al.*, "Loading experiment and thermal analysis for conduction cooled magnet of SMES system," *Front. Electr. Electron. Eng. China*, vol. 4, no. 2, pp. 214–219, 2009.

45. Y. S. Choi, D. L. Kim, and D. W. Shin, "Cool-down characteristic of conduction-cooled superconducting magnet by a cryocooler," *Phys. C Supercond. Appl.*, vol. 471, no. 21–22, pp. 1440–1444, 2011.

46. Y. Iwasa, *Case Studies in Superconducting Magnets Design and Operational Issues*, Second edition, Springer Verlag, Heidelberg, Germany, vol. 53, no. 9, 2013.

47. P. TIXADOR, "Superconducting magnetic energy storage (SMES) systems," in *High Temperature Superconductors (HTS) for Energy Applications*, Woodhead Publishing, Cambridge, UK, 2011, pp. 294–319.

48. D. Sutanto and K. W. E. Cheng, "Superconducting magnetic energy storage systems for power system applications," in *2009 International Conference on Applied Superconductivity and Electromagnetic Devices, ASEMD 2009*, Chengdu, China, September 25–27 2009, pp. 377–380.

5

Superconducting Cables

Rahul Agarwal and Raja Sekhar Dondapati

CONTENTS

5.1 Introduction

The thrust to meet the global energy demand with the growing population and to reduce the depletion of copper mines along with the technological advancements has initiated a strategic development to look out for an alternative of conventional copper cables. In addition, the lifetime of such cables is limited due to the associated *joule loss* in the conductor, which increases the maximum tolerable temperature of its various components, thereby reducing its current transmission capacity. Considering the shortcomings of conventional cables, high-temperature superconducting (HTS) cables are gaining widespread attention as an alternative power transmission device [1].

The inadequate power and aging are posing the greatest obstructions to the present power grid. In the past decades, an ever-rising customers' demand for improved power quality and reliability was observed. Moreover, the growing digital economy relies on a quick and strategic shift to offer novel solutions to customers. In addition, the cost associated with acquiring rights-of-way and decreasing land in urban estates is further gaining momentum, thereby leading to the need for new technologies, which can maintain the desired power density and offer new ways in reducing the overhead cables.

Before the discussion on the role of superconductors as power transmission media, let us briefly introduce the electric grid and the position superconductors occupy in the grid for power distribution purposes.

Before getting into the complexities of superconductors and its role in the electrical power grid, let us discuss about the general grid system. Electrical power is a basic amenity of modern civilization, which provides us lightning and runs our industries and factories. Figure 5.1 shows the schematic representation from power generation to distribution systems. The electrical grid or "the grid" is an interconnected network that supplies electrical power to the world. Between generation and use, several electrical links are present. Initially, generators convert mechanical energy into electrical power when turbines are fueled by fossil fuels, nuclear power or renewable energy. Complex control network and communications equipment are involved, which are supervised by data acquisition system, local and regional control centers. Moreover, in the past decade, rapid growths in computerized system and related components have been observed, further assisting in rapid monitoring and operating purposes. In the present scenario, numerous ongoing challenges are being faced by the current utilities: reliable maintenance and economical service, along with inevitable disruptions. Nevertheless, these challenges will provide a path for superconductors to enter the field of power transmission as discussed below.

The growth in load: In developing countries, the demand for power is accelerating. Further, in the later sections of the chapter, we shall see how such critical growing challenge for conventional cable can be addressed with superconducting technologies.

Environmental and safety issues: Raising climatic issues, such as dust explosion, have imposed a stress on engineers to reduce the carbon-emitting components. Moreover, the highly flammable oil utilized as coolant in conventional power equipment further raises the risk of oil leakage, which could lead to fire hazards. Such problems can be solved by employing superconducting cables, which utilize non-flammable coolants, such as liquid nitrogen (LN_2).

Power quality and reliability: In the future, high energy density would be an essential requirement, which would be difficult to achieve using copper cables due to the associated *joule heating* effect. Moreover, sudden malfunction or fault in the equipment would further disrupt the working of electrical components. In addition,

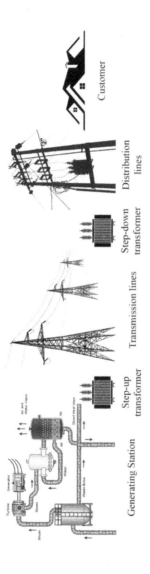

FIGURE 5.1
Schematic representation of the power generation and transmission system.

over-currents in the grid would increase the operating temperature of the component. Hence, to overcome such issues, superconducting cables are becoming a popular choice for incorporation into futuristic grids.

At this stage, a quantitative comparison between the conventional cable and the superconducting cable would be beneficial in order to brief the reader regarding the power transmission capacity of both the devices. Table 5.1 summarizes the difference between the conventional cable and warm and cold dielectric HTS cables [2]. In the later section, we will introduce the concept of warm dielectric (WD) and cold dielectric (CD) HTS cables. From comparison, it is evident that for the same applied voltage, WD and CD HTS cables have more power transmission capacities compared to that of conventional cable and the electrical losses are also reduced for CD HTS cable.

In these contexts, the infiltration of superconducting equipment in the grid would enhance performance, which in many cases would enable to replace the current conventional technology. The HTS cable has emerged as a promising technology to address these issues. Figure 5.2 shows the role of superconducting cables for power transmission. These cables can replace the conventional cables with significant reduction in the space required for conventional cables. Listed below are the key benefits of employing HTS cables:

- Current handling capacity of HTS cables is three to five times that of a conventional cable,
- Minimum waster heat or electrical losses,
- Feasibility in installation in densely populated regions due to less space consumption than conventional cables,
- High-power carrying capacity at lower voltages, which would further lead to the elimination of one or more transformers and associated equipment,
- Effective control of power flow across meshed grids,
- Usage of environment friendly coolants, such as LN2 for cooling purposes,
- Lower impedance than conventional cables and overhead lines [3].

TABLE 5.1

Comparison of Power Capacity for Different Cable Designs

Technology	Pipe Outer Diameter (inch)	Voltage (kV)	Power (MVA)	Losses (W/MVA)
Conventional cable	8	115	220	300
Warm dielectric HTS cable	8	115	500	300
Cold dielectric HTS cable	8	115	1000	200

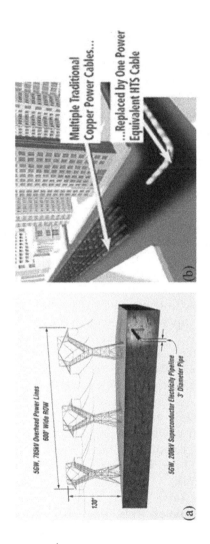

FIGURE 5.2
Role of superconducting cables in replacing the existing (a) grid system and (b) distribution cables for the same power rating.

Figure 5.3 show a schematic representation of the HTS cable. HTS tapes are spirally wrapped around a former (cooling channel), enclosed in a cryostat. A detailed description of the HTS cable is provided in Section 5.3. In the past decades, power generation and distributions at medium to high voltages have been recognized as a significant area for superconductors to contribute. Following several demonstrations across the globe, cables fabricated using high-temperature superconductors are closest to full commercialization. Garwin and Matisoo [4] conducted the first detailed study in which they emphasized that low-temperature conductors based on superconducting cable cooled by liquid helium would have to carry many gigawatts of power to be commercially feasible. The most significant demonstration of AC cable based on low-temperature superconductors was conducted at Brookhaven National Laboratory, started in 1972. This cable was fabricated using Nb_3Sn tapes and cooled using liquid helium and was designed to transport 1 GW of three-phase power with three such cables at 138 kV [5]. However, the cost of the system and the complexities involved in helium cryogenics prevented the infiltration of such cables in power grid. Such events made liquid helium–cooled superconducting cables viable in controlled environment of high-energy physics laboratory. In the next years, superconductivity of some other metals, alloys and intermettalic compounds has also been discovered, as shown in Figure 5.4; however, the demonstration of superconductivity was at very low critical temperature (T_c = 23.2 K for Nb_3Ge). Such circumstances delayed the practical applications of superconductors due to high cost of helium (~ \$25 per liter), in addition to its preparation difficulties.

Such a situation changed dramatically in 1986 with the discovery of the so-called high-temperature superconductivity in non-traditional compounds, namely cuprates. These compounds operate at the temperature range of liquid nitrogen (LN_2), which simplified the refrigeration design and cost, reducing the cryogenic load. Moreover, the availability of robust and high-critical current first generation (1G) BSCCO-2223 HTS tapes in mid-1990s, HTS cable projects were launched around the world as shown in Table 5.2. These cables typically contain HTS wires and a wrapped dielectric, similar to conventional AC cables; however, sophistication arises when LN_2-cooled superconductors are mated with dielectric at room temperature. Moreover, such systems are accompanied with refrigeration equipment, cryocoolers, pumps and LN_2 storage tank.

These projects grew in sophistication, along with the increasing current carrying capacity. Unfortunately, this progress was not always steady. A HTS cable project fabricated at Detroit Edison by Pirelli Cables and Systems [28] failed in 2001 due to the failure of vacuum jackets. However, Sumitomo Electric Industries (SEI) and Furukawa Electric Company, Ultera and a collaboration of Southwire and NKT cables continued their programs [10, 12, 18]. Moreover, they were joined by Nexans, which supported the Long Island Power Authority (LIPA) cable project with American Superconductors (AMSC) and successfully demonstrated the first in-grid 138 kV on Long

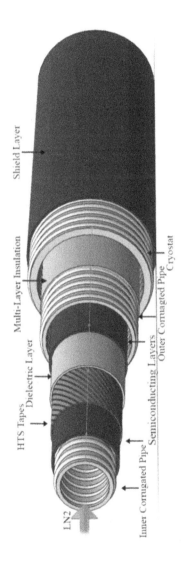

FIGURE 5.3

Schematic view of the HTS cable.

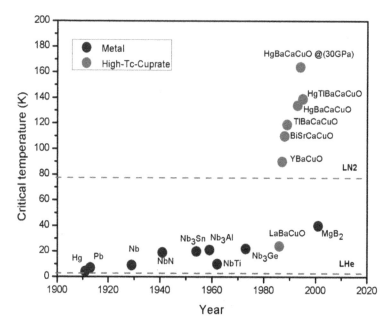

FIGURE 5.4
Discovery of superconductor with increasing T_c.

Island in 2008 [29]. Programs were started in China and Korea with the demonstration of first in-grid cables utilizing second-generation (2G) HTS wire in 2011 [24]. Furthermore, programs continued in Russia as well, with the development of both AC and DC HTS cables [17]. One of the most significant in-grid HTS cable demonstration is the AmpaCity project in Essen, Germany, where a 1 km long 10 kV/40MVA distribution cable is developed, along with a standalone fault current limiter, which links two substations and thus protecting a 110 kV cable [26].

With the development of the HTS cable, the most obvious benefit of the reduction in electrical loss is still under development. Issues such as hike in cryogenic refrigeration load due to superconductor AC loss, dielectric losses, and thermal losses, largely nullify its inherent advantage. However, a significant cost incurred with the inclusion of extra HTS conductor, multi-layer insulation and more efficient turbo-Brayton refrigerators. Moreover, the conventional cables could be fabricated with more copper with the trade-off between weight, diameter and cost. In these aspects, AC HTS cables have the potential to transfer larger current densities at a similar power with much lower voltage. Such capacity allows to replace the underground ducts with higher power HTS cables, without creating additional complexities, thereby making its installation easy. Moreover, inherent characteristics of non-linear shifting of electrical resistance, when one or more than one parameter(s) retaining superconductivity are violated. Such change in electrical resistance

TABLE 5.2

Major World-Wide Projects of the AC HTS Cable

Type	Cable Manufacturer	Rating	Wire manufacturer	References
WD-1 core	Pirelli	115 kV/2 kA/50 m/1Φ	BCSSO-AMSC	[6]
	Innost/Innopower	35 kV/2 kA/33 m/3Φ	BSCCO-Innost	[7]
	ChangTong	10.5 kV/1.5 kA/75 m/3Φ	BSCCO-AMSC	[8]
	Ultera	13 kV/3 kA/200 m/3Φ	BSCCO-AMSC	[9]
WD-Tri-axial	NKT cables	30 kV/2 kA/30 m/3Φ	BSCCO-NST	[10]
CD-1 core	Southwire	12.4 kV/1.25 kA/30 m/3Φ	BSCCO-AMSC	[11]
	Furukawa	77 kV/0.7 kA/30 m/1 Φ	BSCCO-Furukawa	[12]
	Furukawa	77 kV/1 kA/500 m/1Φ	BSCCO-Furukawa	[13]
	Nexans	138 kV/1.8 kA	BSCCO-AMSC	[14]
	Nexans	20 kV/3.2 kA/30 m/1Φ	BSCCO-AMSC	[15]
	Furukawa	275 kV/3 kA/30 m/1Φ	YBCO-AMSC	[16]
	VNIIKP	20 kV/2 kA/200 m/3Φ	BSCCO-SEI	[17]
CD-3 core	SEI	66 kV/1 kA/100 m/3Φ	BSCCO-SEI	[18]
	LG Cable	23 kV/1.26 kA/30 m/3Φ	BSCCO	[19]
	SEI	34.5 kV/0.8 kA/30 m/3Φ	BSCCO-SEI/YBCO-SP	[20]
	SEI	23 kV/1.25 kA/100 m/3Φ	BSCCO-SEI	[21]
	LS Cable	23 kV/1.25 kA/100 m/3Φ	BSCCO-AMSC	[22]
	SEI	66 kV/1.75 kA/30 m/3Φ	BSCCO-SEI	[23]
	LS Cable	23 kV/1.25 kA/500 m/3 Φ	YBCO-AMSC	[24]
	SEI	66 kV/5 kA/15 m/3Φ	YBCO-SEI	[16]
	SEI	66 kV/1.75 kA/250 m/3Φ	BSCCO-SEI	[25]
CD-Tri-axial	Nexans	10 kV/2.3 kA/1000 m/3Φ	BSCCO-SEI	[26]
	Ultera	10 kV/4 kA/300 m/3Φ	YBCO-AMSC	[27]

is beyond the scope of normal copper conductor, which enables the development of fault-current limiting HTS cables [30]. Such functionality provides an opportunity to directly connect substations, eliminating the potentially catastrophic fault currents between substations. Moreover, the installation of HTS cables would be beneficial for high-population density regions. Compared to the AC HTS cable, DC cables would be a better mean to reduced losses, due to exclusion of AC losses. However, the existing power grids employ AC and any DC link must be converted to AC through expensive electronic systems. In a nutshell, the successful demonstration of in-grid HTS cable demonstration internationally has brought superconducting cable technology to a significant technical level to utilize at the utilization end.

5.2 Configurations of High-Temperature Superconducting Cables

Several designs of cables have been prototyped and developed in order to increase the efficiency of superconducting systems, while minimizing the operational cost arising due to the cost of HTS materials and refrigeration system. The variation in cable geometry affects the efficiency of the HTS cable and thus can be classified on the basis of dielectrics as shown in the. At present, there are two principal types of HTS cables, as shown in Figure 5.5. In the first design, the single superconducting layer consisting of HTS tapes is wrapped around a flexible core. This cable design employs an outer dielectric layer (electric insulation) which restricts the flow of current in radial direction. This insulation layer is present at room temperature and is called "warm dielectric" design (Figure 5.6(a)). The cable assembly is contained within a thermal insulation for preventing any heat leakage into the HTS core. The whole HTS core is inserted into a cryostat. Such configuration offers high power density and minimum amount of superconducting material. However, drawbacks are associated with such configuration, such as

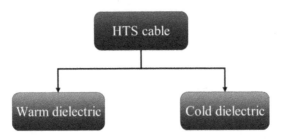

FIGURE 5.5
Classification of HTS cable based on warm and cold dielectric cables.

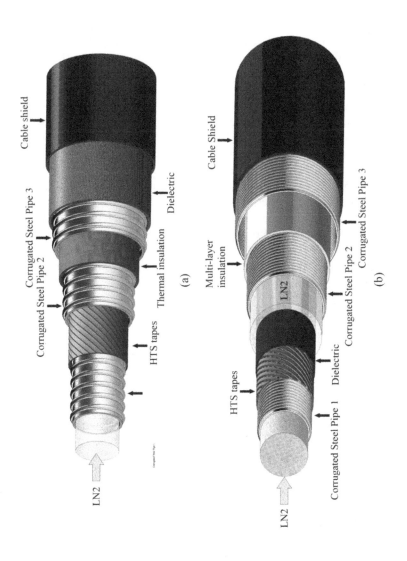

FIGURE 5.6
Schematic representation of the configurations of HTS cables: (a) warm dielectric and (b) cold dielectric cables.

high electrical losses (increasing refrigeration cost) and higher impedance. Warm dielectric (WD) design of the HTS cable is most suitable for retrofitting in existing pipe or duct systems. A WD HTS cable has the potential to transport twice the power of a conventional cable with approximately similar losses. The design of the WD HTS cable has many characteristics, which make it attractive for the near future. The absence of a superconducting shield layer results in a lower initial cost. Moreover, the accessories could be derived from conventional designs and has similar handling procedures as a conventional cable. In the cryostat, a vacuum region with radiative multi-layer insulation is provided in order to prevent heat-in-leak from the surrounding.

Another configuration is a cold dielectric cable, which as HTS tapes spirally wrapped around a flexible corrugated pipe. A dielectric layer in applied over the HTS tapes and a second layer of HTS tape is applied over the insulation. LN_2 flows over and between the layers, thereby providing dielectric characteristics; such configuration is commonly referred to as the "cold dielectric" design (Figure 5.6(b)). Such design utilizes maximum amount of HTS tapes but it offers several advantages, such as reduced AC losses, higher current carrying capacity and complete suppression of stray electromagnetic field outside the cable assembly. The reduction of AC losses further aids in the extension of HTS cables and wider spacing of the cooling stations. Moreover, the inductance of the cold dielectric HTS cable is significantly lower than a conventional cable. A cold dielectric HTS cable has the greatest level of transmission of transmission capacity and efficiency.

5.3 Design Analysis of High-Temperature Superconducting Cables

The design of any superconducting device requires the consideration of two major aspects: (1) vacuum insulation and (2) cryogenics, as shown in Figure 5.7. The presence of vacuum insulation eliminates the infiltration of heat transfer via conducting or convection. In the absence of vacuum, the infiltration of heat from ambient would cause the transition of material from the superconducting state to the normal conducting state. Hence, vacuum insulation forms the primary segment of any superconducting device. Following it is the chill-down operation using cryogens, such as LN_2. For the material to be a super conductor, its temperature is required to be brought down below its critical temperature. Such cooling is achieved either by forced convection [31] or conduction [32].

The first step for the fabrication of HTS cables requires the determination of the current to be transported. For this process, a selection of the available superconducting tape is required. A superconductor has three parameters

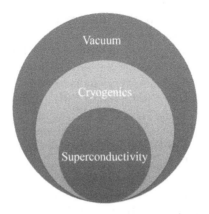

FIGURE 5.7
Primary aspects required for the construction of superconducting devices.

under which superconductivity is attained. These parameters are *critical temperature* (T_c), *critical external magnetic field* (H_c) and *critical current density* (J_c). These critical parameters are not independent of each other; rather they are in strong correlation with each other. Figure 5.8 shows the relation between the three parameters. Any point enclosed by the surface between the planes (J_c, H_c), (H_c, T_c) and (T_c, J_c) would be in the superconducting state; any point lying on the surface of (J_c, H_c, T_c) would be in the critical state and the point lying outside the surface would be in the normal conducting state. Further, for any superconductor the volume enclosed between the

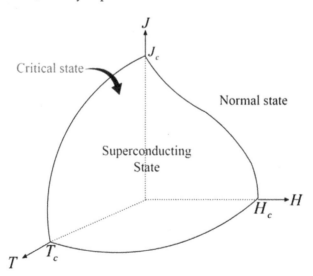

FIGURE 5.8
Critical parameters – current (J_c), temperature (T_c) and external magnetic field (H_c) and their relationships for a superconductor.

three axes of J_c, H_c and T_c is constant. Hence, a superconductor with larger J_c would also suffer to a reduction in H_c and T_c. Various superconducting tapes are available commercially and based on the targeted application the tape should be selected. For example, when the superconducting cable is to used as magnet grade, the primary focus is to have larger H_c and J_c, where higher critical temperature is not a major concern. Hence, a reduction in T_c would subsequently increase H_c and J_c. Hence, the reader should select the tape having higher margin for H_c and J_c, whereas superconducting cables to be employed for power transmission purposes for electric grid should have higher margin for J_c and H_c.

The fundamental interest of superconductivity in electric power applications arises due to the absence of ohmic resistance below J_c, H_c, and T_c, depending on the superconducting material. Figure 5.9 shows a typical voltage-current characteristic of a superconducting tape. A singular point on this curve is the Tc, which is conventionally estimated at 1 mV/cm electric field along the wire. Below this critical current, the electrical resistance can be neglected and the material attains superconducting state. Above this value, the material rapidly loses its superconducting properties and becomes resistive. Such transition is expressed by the E–J power law, given by:

$$E(T) = E_c \left(\frac{J_{op}}{J_c(T)} \right)^n \qquad (5.1)$$

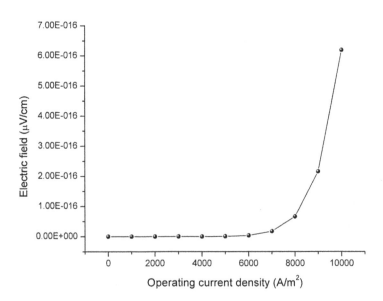

FIGURE 5.9
E-J characteristic of a superconductor.

where n is the superconducting transition index, T is the absolute temperature and E_c is the electric field defining the critical current (1 µV/cm). Figure 5.10 represents the steps involved for the design of HTS cable. Hence, in the coming sections, we would discuss the construction details of HTS cable.

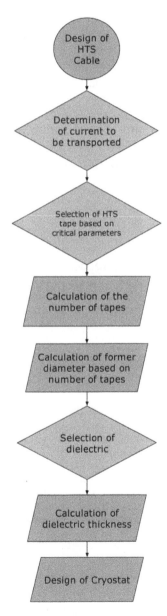

FIGURE 5.10
Steps involved in the design of the HTS cable.

5.3.1 Superconducting Materials for HTS Cables

In HTS cable, superconductors form the media for the transport of current, thus forming the most critical part of cable system design. When the design of superconducting cable is concern, one should account for various parameters in current carrying conductors in order to form a replaceable system for the existing copper cables. These parameters consists of the availability of materials for long lengths, flexibility to during the winding and installation operation, and able to withstand thermal contraction during chilldown operation. Till this date, three kinds of HTS materials possess these characteristics. The first two materials are available in the form of tapes, $Bi_2Sr_2Ca_2Cu_3O_{10-x}$ (Bi-2223) (first generation conductors) with a $T_c \sim 108$ K and $YBa_2Cu_3O_{7-x}$ (YBCO) (second generation or coated conductors) with a $T_c \sim 92$ K. The third one is in the form of cylindrical wire, which is based on MgB_2 and is superconducting below 39 K. Figure 5.11 represents the details of the two tapes and wire architecture, and their performances are mentioned in Table 5.3.

5.3.2 Number of Tapes

After the selection of HTS tapes, the number of tapes is to be decided. Suppose say, the current carrying capacity for a HTS tape is 100 A and the required current to be transported through HTS cable is 1.2 kA. Hence, the number of tapes required would be 12.

5.3.3 Former Diameter

The determination of the number of tapes would allow selecting the former diameter, given by the relation,

$$2\pi r = n_{\text{tape}} w_{\text{tape}} \tag{5.2}$$

$$r = \frac{n_{\text{tape}} w_{\text{tape}}}{2\pi} \tag{5.3}$$

where r is the outer radius of former, n_{tape} is the number of HTS tapes, and w_{tape} is the width of tape, as shown in Figure 5.12.

5.3.4 Dielectric Insulation

A cable with dielectric tape is typically about one inch wide. Measurements were normally made on prototype cable, where the insulation comprises of several layers of dielectric tapes applied to the cable. Many dielectric materials were tested between 1972 and 1974 under a US-DOE program using 10 ft long samples with a 3 ft test section. The breakdown strength

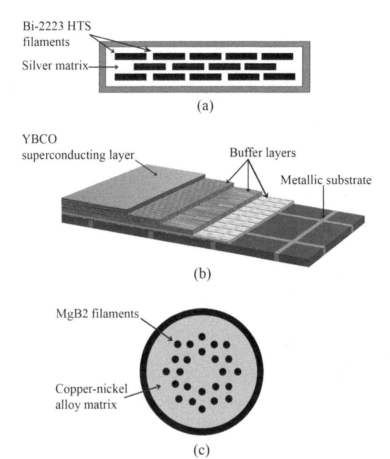

FIGURE 5.11
Structural details of HTS tapes and wires: (a) BSCCO tape, (b) YBCO tape and (c) MgB$_2$ wire.

TABLE 5.3

Properties of Superconducting Tapes and Wires

		Dimensions		Commercial Performances	
	Shape	Width (mm)	Thickness (mm)	J_c (A/mm^2) at 77 K, self-field (A/cm)	Length (m)
Bi-2223	Laminated tapes	4, 5	0.3	120–150	< 1500
YBCO	Laminated NiW tapes	4–12	0.2	100–150	< 500
	Ion beam assisted deposition (IBAD) tapes		0.05	400–500	
	Shape	Diameter (mm)		J_c (A/mm^2) at 20 K, 1 T	Length (m)
MgB$_2$	Cylindrical wires	0.8–1.5		300	< 3000

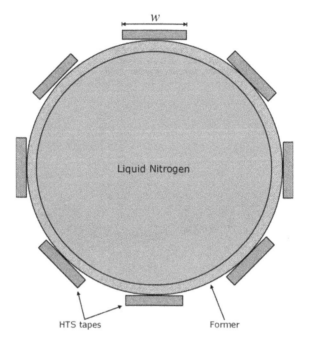

FIGURE 5.12
Schematic cross section of a HTS cable with HTS tapes placed on the corrugated pipe (former).

was determined to be 201 kV in a 0.22 inch thick sample at a test length of 36 inches. The peak and the average breakdown strength of this sample was 1099 and 913 kV/inch respectively. Measurements were conducted on various built-up thicknesses, which helped in the development of scaling function given by equation (5.4). For a cable of insulation thickness of about 0.8 inch (=20 mm), the peak strength was projected to be around 900 kV/ inch (36 kV/mm). The constant A and B in equation (5.4) are estimated to be 47.8 kV/mm and 0.1 respectively from the measured data. However, this equation is valid for insulation thickness less than 50 mm:

$$E_m = At_{ins}^{-B} \tag{5.4}$$

where, E_m is the peak electric stress (kV/mm), tins represents insulation thickness (mm), A = 47.8 kV/mm, and B = 0.1.

Figure 5.13 represents a generic cable configuration with conducting and insulation. The thickness of insulation can be determined using maximum working stress. The peak stress (E_m) generally occurs at the surface of the cable conductor and is given by,

$$E_m = \frac{V}{R_1 \ln\left(\dfrac{R_2}{R_1}\right)} \tag{5.5}$$

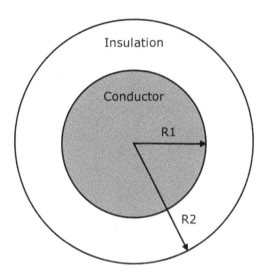

FIGURE 5.13
Cross section of the HTS core.

$$t_{ins} = R_1 \left(e^{\frac{V}{E_m R_1}} - 1 \right) \tag{5.6}$$

where E_m is the peak dielectric stress at conductor surface, R_1 and R_2 are the inside and outside radius of insulation (outer radius of conductor layer) respectively and V is the conductor voltage.

The equation (5.5) can be rearranged to provide the radial thickness of (t_{ins}) of insulation. Generally, the peak stress (E_m) is selected as half to a third of the breakdown strength of the dielectric material. Moreover, due to the non-uniformity of the outer surface, a general practice it to incorporate few layers of a carbon-black paper, which is made from cellulose or polyethylene paper embedded with carbon particles.

5.3.5 Design of Cryostat

The cryogenic environment for HTS cable core is provided by the cryostat, shown in Figure 5.14. It consists of an inner cold wall and an outer warm wall. The walls consist of corrugated stainless steel pipe, which facilitate thermal contraction of HTS cable during chill-down process. The space between the two walls is separated by vacuum, along with multi-layer insulation (MLI). The cryostat is subjected to various thermal loads, which includes thermal conduction through cryostat walls from room temperature to LN$_2$ temperature, convection thermal load by gas molecules in the space between the inner cold and outer warm walls, and radiation heat transfer. The presence of vacuum helps reducing the

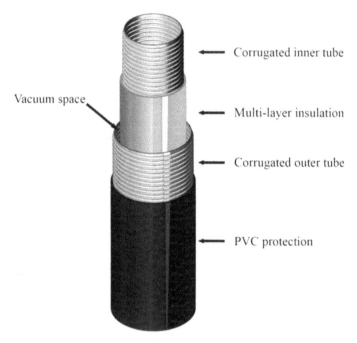

FIGURE 5.14
Schematic representation of a superconducting cable cryostat.

convective heat transfer, which arises from the heat conducted from the surrounding. Moreover, the presence of MLI further reduces the radiation heat transfer.

5.3.6 Cooling Strategies

One of the most important components of HTS cable system is the cooling facility for achieving sufficiently low temperature for superconductors to maintain superconductivity. During the operating condition of HTS DC/AC cable, the coolant must absorb the thermal heat and AC losses. In the present scenario, two types of cooling arrangement exist, as shown in Figure 5.15. In order to save space, it is desirable to return coolant within the cable itself, rather than having a separate return pipe. For long length cable fabrication, counter-flow cooling posses some issues. If the thermal conductivity of dielectric is sufficiently large, the temperature of the returning coolant may exceed the temperature of the outer coolant, or in worst case may exceed the critical temperature of superconductors. Hence, such configuration will be suitable for relatively short HTS cable. Hence, the cooling configuration with single flowing method is a good candidate for long HTS cable.

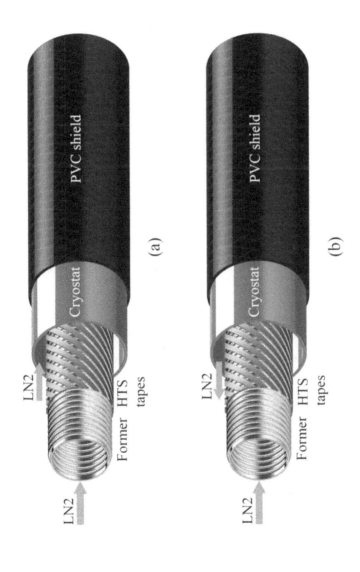

FIGURE 5.15
Cooling configurations of HTS cable: (a) parallel flow and (b) counter flow cooling.

5.3.7 Pumping Losses and Temperature Difference

The formation of velocity gradient across the cross-section of the LN_2 transfer line and the cryostat leads to utilization of pumping power. In the turbulent regime, due to corrugations of the transfer line, the Reynolds number of described by:

$$\text{Re} = \frac{\rho v D_h}{\mu} \tag{5.7}$$

The pressure drop and temperature difference due to circulation of LN_2 is expressed as:

$$\Delta P = \frac{f \rho v^2 L}{2 D_h} \tag{5.8}$$

$$\Delta T = \frac{Q}{m c_p} \tag{5.9}$$

where f, ρ, v, m, c_p represents friction factor, density, velocity, mass flow rate and specific heat of LN_2 respectively. D_h represents equivalent curved surface area and hydraulic diameter of transfer lines and Q represents heat loss in the cable.

5.4 Existing Challenges and Opportunities

Since the 1960s, underground transmission cables with cross-linked polyethylene (XLPE) insulation have been in use worldwide. These cables can be built where there is a lack in space, thus reinforcing the existing transmission system where necessary. Moreover, such cables are not vulnerable to ice or wind storm, sagging due to overload and similar environmental factors. Hence, the biggest challenge for HTS cable is to compete with these maintenance-free and reliable cable systems. From the user's perspective, low cost along with non-diminished reliability are the criteria for product selection. Some ongoing challenges faced by the HTS cable are described below.

5.4.1 Large Concentration of LN_2

Nitrogen is a colorless, odorless and non-flammable gas, which occurs naturally in the atmosphere. When cooled, nitrogen condenses into a liquid at a boiling point of 77 K and is abundant in nature. Hence, nitrogen is an ideal coolant for electric power applications in HTS cables. However, there are two main risks associated with nitrogen: (1) oxygen deficiency and (2) extreme low temperatures. The released amount of high concentration nitrogen in confined spaces can lead to asphyxiation due to the displacement of oxygen. Further, raised concentration of nitrogen may cause a variety of respiratory symptoms. If a leak or spill occurs, nitrogen will rapidly dissipate into the environment. Hence, in order for the HTS cable to be adapted on a wide-spread basis, there are several considerations to be achieved, such as instal-lation of a system of sensors along the length of HTS cable to detect low temperature or the concentration of oxygen. Moreover, monitoring of coolant flow, temperature and pressure can be used to determine leakage.

5.4.2 Capital and Operating Cost

The cost of the HTS wire, the cryostat and the involved refrigeration is cur-rently high. The capital cost of the HTS system could be three to four times that of a conventional system. Despite the compactness achieved in the HTS cable, a major drawback to its application in electric grid is the capital cost.

5.4.3 Refrigeration and Cooling System Reliability

The cooling system of a HTS cable is highly critical. The failure of the HTS cable system could lead to superconductors carrying the nominal load for a short time. Thus, the cooling system for HTS cables must be highly reli-able. In order to achieve the desired cooling power, several cryocoolers could be combined. Hence, the necessity to maintain the cooling require-ment presents the failure mode for HTS cables. The cooling system is the most critical component of the HTS cable system because the cable is unable to carry any significant load unless it operates close to the nominal operating temperature. Thus the necessity to maintain cooling at cryo-genic (liquid nitrogen) temperatures presents a failure mode for the HTS cable system.

5.4.4 Performance during Normal and Fault Currents

During a fault experienced by the HTS cable, the HTS tapes are subjected to quench. Once quenching occurs, *joule heating* raises the temperature of the HTS tapes, which forces a shutdown of the cable system. Moreover, such quenching could create a hot spot, which could lead to cable burnout. An immediate restoration of the HTS cables after experiencing a high current fault cannot be achieved. Moreover, higher voltage cables are more difficult to cool down due to the thick covering of dielectric layer, which has poor thermal conduction. Hence, contingency backups would be required in such situations. Further, the HTS cable takes a longer time to cool down from room temperature to their operating temperature. After experiencing a fault, these cables warm up to cause a shutdown for several hours. Under such circumstances, alternative routes must be available to carry the load of the HTS cable.

5.4.5 FCL Cable

Fault current limiting capability in the HTS cable is proposed to be incorporated. The FCL cable employs a flux flow resistive region of the HTS tape when the wire resistance goes up. Such increased resistivity could be utilized for limiting the fault current. However, a uniform transition from the superconducting region to the flux flow region is not guaranteed. Hence, the performance of such cable in an electric grid is not clear. Therefore, a critical analysis of the FCL cable is required before integration into the grid. For example, under fault conditions, it is a fundamental property of the HTS conductor to have an immediate transition to normal state, once the current exceeds the critical current. Such transformation can occur within milli-seconds. Then all current is switched to the copper core (and copper tapes if necessary). Once a cable quenches, the cable requires a longer time to cool down owing to its operating temperature.

5.5 Summary

Cables employing LN_2 as coolant include a vacuum-insulated cryostat. The operating cost of refrigeration and thermal loads is the important component in the economic evaluation of a HTS cable. The high component cost makes such cables economically feasible only when high power is imposed. However, the losses pertaining to the present grid and a wide blackout are still a major concern, for which the adaptation of the HTS cable is a necessity in the near future. Moreover, an initial application of HTS technology is expected to be in crowded urban region where space limitation is a major concern. However, the biggest challenge for a HTS cable system would be to compete with the highly reliable XLPE cable system.

References

1. R. Wesche, A. Anghel, B. Jakob, G. Pasztor, R. Schindler, and G. Vécsey, "Design of superconducting power cables," *Cryogenics (Guildf).*, vol. 39, no. 9, pp. 767–775, 1999.
2. J. Jipping, A. Mansoldo, and C. Wakefield, "The impact of HTS cables on power flow distribution and short-circuit currents within a meshed network," in *2001 IEEE/PES Transmission and Distribution Conference and Exposition. Developing New Perspectives (Cat. No.01CH37294)*, 2001, vol. 2, pp. 736–741.
3. J. Howe, B. Kehrli, F. Schmidt, M. Gouge, S. Isojima, and D. Lindsay, "Very low impedance (VLI) superconductor cables: Concepts, operational implications and financial benefits," *Whitepaper, Am. Supercond. Corp.*, November, pp. 1–29 2003.

4. R. L. Garwin and J. Matisoo, "Superconducting lines for the transmission of large amounts of electrical power over great distances," *Proc. IEEE*, vol. 55, no. 4, pp. 538–548, 1967.

5. E. B. Forsyth and R. A. Thomas, "Performance summary of the brookhaven superconducting power transmission system," *Cryogenics (Guildf).*, vol. 26, no. 11, pp. 599–614, 1986.

6. M. Nassi, N. Kelley, P. Ladié, P. Corsato, G. Coletta, and D. Dollen, "Qualification results of a 50 m-115 kV warm dielectric cable system," *Appl. Supercond. IEEE Trans.*, vol. 11, pp. 2355–2358, 2001.

7. Y. Xin *et al.*, "Introduction of China's first live grid installed HTS power cable system," *IEEE Trans. Appl. Supercond.*, vol. 15, no. 2, pp. 1814–1817, 2005.

8. L. Y. Xiao *et al.*, "Development of HTS AC power transmission cables," *IEEE Trans. Appl. Supercond.*, vol. 17, no. 2, pp. 1652–1655, 2007.

9. J. A. Demko *et al.*, "Triaxial HTS cable for the AEP bixby project," *IEEE Trans. Appl. Supercond.*, vol. 17, no. 2, pp. 2047–2050, 2007.

10. O. Tønnesen *et al.*, "Operation experiences of a 30 kV/100 MVA high temperature superconducting cable system," *Supercond. Sci. Technol.*, vol. 17, p. S101, Feb. 2004.

11. J. P. Stovall *et al.*, "Installation and operation of the southwire 30-meter high-temperature superconducting power cable," *IEEE Trans. Appl. Supercond.*, vol. 11, no. 1, pp. 2467–2472, 2001.

12. S. Mukoyama, H. Hirano, M. Yagi, H. Kimura, and A. Kikuchi, "Test results of a 30 m high-Tc superconducting power cable," *IEEE Trans. Appl. Supercond.*, vol. 13, no. 2, pp. 1926–1929, 2003.

13. T. Takahashi *et al.*, "Demonstration and verification tests of 500 m long HTS power cable," *IEEE Trans. Appl. Supercond.*, vol. 15, no. 2, pp. 1823–1826, 2005.

14. J. F. Maguire *et al.*, "Development and demonstration of a HTS power cable to operate in the long island power authority transmission grid," *IEEE Trans. Appl. Supercond.*, vol. 17, no. 2, pp. 2034–2037, 2007.

15. R. Soika, X. G. Garcia, and S. C. Nogales, "ENDESA supercable, a 3.2 kA, 138 MVA, medium voltage superconducting power cable," *IEEE Trans. Appl. Supercond.*, vol. 21, no. 3, pp. 972–975, 2011.

16. O. Maruyama *et al.*, "Development of 66 kV and 275 kV class REBCO HTS power cables," *IEEE Trans. Appl. Supercond.*, vol. 23, no. 3, p. 5401405, 2013.

17. E. P. Volkov, V. S. Vysotsky, and V. P. Firsov, "First Russian long length HTS power cable," *Phys. C Supercond. Appl.*, vol. 482, pp. 87–91, 2012.

18. S. Honjo *et al.*, "Electric properties of a 66 kV 3-core superconducting power cable system," *IEEE Trans. Appl. Supercond.*, vol. 13, no. 2, pp. 1952–1955, 2003.

19. D.-W. Kim *et al.*, "Development of the 22.9-kV class HTS power cable in LG cable," *IEEE Trans. Appl. Supercond.*, vol. 15, no. 2, pp. 1723–1726, 2005.

20. T. Masuda *et al.*, "Fabrication and installation results for Albany HTS cable," *IEEE Trans. Appl. Supercond.*, vol. 17, no. 2, pp. 1648–1651, 2007.

21. S. H. Sohn *et al.*, "The results of installation and preliminary test of 22.9 kV, 50 MVA, 100 m Class HTS power cable system at KEPCO," *IEEE Trans. Appl. Supercond.*, vol. 17, no. 2, pp. 2043–2046, 2007.

22. K. Sim *et al.*, "DC critical current and AC loss measurement of the 100m 22.9kV/50MVA HTS cable," *Phys. C Supercond.*, vol. 468, no. 15, pp. 2018–2022, 2008.

23. T. Masuda *et al.*, "Test results of a 30 m HTS cable for Yokohama Project," *IEEE Trans. Appl. Supercond.*, vol. 21, no. 3, pp. 1030–1033, 2011.

24. S. Lee, J. Yoon, B. Lee, and B. Yang, "Modeling of a 22.9kV 50MVA superconducting power cable based on PSCAD/EMTDC for application to the Icheon substation in Korea," *Phys. C Supercond. Appl.*, vol. 471, no. 21, pp. 1283–1289, 2011.

25. H. Yumura *et al.*, "Update of YOKOHAMA HTS cable project," *IEEE Trans. Appl. Supercond.*, vol. 23, no. 3, p. 5402306, 2013.

26. M. Stemmle, F. Merschel, M. Noe, and A. Hobl, "AmpaCity — Installation of advanced superconducting 10 kV system in city center replaces conventional 110 kV cables," in *2013 IEEE International Conference on Applied Superconductivity and Electromagnetic Devices*, 2013, 25–27 October, 2013, Beijing, China, pp. 323–326.

27. J. Maguire *et al.*, "Status and Progress of a Fault Current Limiting HTS Cable to be Installed in the Con Edison Grid," *AIP Conf. Proc.*, vol. 1218, no. 1, pp. 445–452, 2010.

28. N. Steve, M. Nassi, M. Bechis, P. Ladiè, N. Kelley, and C. Wakefield, "High temperature superconducting cable field demonstration at Detroit Edison," *Phys. C Supercond.*, vol. 354, no. 1, pp. 49–54, 2001.

29. J. F. Maguire *et al.*, "Progress and status of a 2G HTS power cable to be installed in the long island power authority (LIPA) grid," *IEEE Trans. Appl. Supercond.*, vol. 21, no. 3, pp. 961–966, 2011.

30. H. Kojima, F. Kato, N. Hayakawa, M. Hanai, and H. Okubo, "Superconducting fault current limiting cable (SFCLC) with current limitation and recovery function," *Phys. Procedia*, vol. 36, pp. 1296–1300, 2012.

31. R. S. Dondapati, G. S. Member, and V. V Rao, "CFD analysis of cable-in-conduit conductors (CICC) for fusion grade magnets," *IEEE Trans. Appl. Supercond.*, vol. 22, no. 3, pp. 4703105–4703108, 2012.

32. Y. S. Choi, D. L. Kim, and D. W. Shin, "Cool-down characteristic of conduction-cooled superconducting magnet by a cryocooler," *Phys. C Supercond. Appl.*, vol. 471, no. 21, pp. 1440–1444, 2011.

6

Superconducting Fault Current Limiters

Rajesh Kumar Gadekula and Raja Sekhar Dondapati

CONTENTS

6.1 Introduction

Electrical faults in engineering systems are attributable to

a) Weather conditions such as lightning, heavy winds and rains, accumulation of snow and ice and the deposition of salt on overhead lines and conductors.

b) Failures in equipment such as transformers, generators, reactors, motors, cables and windings of insulation etc.

c) Human-based errors such as rating the equipment improperly, during service switching circuit, uneven maintenance etc.

d) The presence of smoke, which ionizes air particles, resulting in sparks in the overhead lines. Hence, the insulating capability at higher voltages will be decreased.

These faults cause over flow in current, hazardous conditions in grid and disturbance in interconnected circuits thereby costlier equipment will be damaged.

The electrical power grids are protected with the conventional switchgear equipment, circuit breakers, electromagnetic relays and few other protection devices to protect the power devices from the electrical failures. However, these equipment should be replaced and monitored after the fault operation. Superconducting Fault Current Limiters (SFCLs) have transformed the way that electrical power systems handle fault currents within milli-seconds without interruption in the power demand. These include having better performance and mitigation of fault current within the first cycle of fault compared with conventional protection devices. The commercialization of long length coated conductors (CCs), such as YBCO, with a critical temperature of 90 K at 0 Tesla and BSCCO, with a critical temperature of 110 K at 0 Tesla, made the wide use in the manufacturing of SFCL. The loss-free transmission of electrical power during normal operation, the fault compensation within milli-seconds and the fast recovery from the normal state to the superconducting state after fault compensation are the ideal advantages of SFCLs. Further SFCLs are also termed as unmanned protection devices for power grids. Hence, the demand on commercialization of SFCLs is increasing drastically in today's electrical power systems for protecting the power devices and for safe operation without power interruptions.

6.2 Types of Faults

Steady operations of the power systems are affected by the instability due to the overloads and short circuit and are stated as faults. There are two types of faults in three phase electrical power grids. They are open circuit faults and short circuit faults. The open circuit faults are further classified as single phase, two-phase and three-phase conductors. Further the short circuit faults are classified into symmetrical and unsymmetrical faults. Hence, in this section, the different types of faults that a power system can encounter during operation are discussed in (Figure 6.1).

6.2.1 Open Circuit Faults

These are faults that occur due to the failure of one or more conductors in the power grid. These are also called "series faults" because the conductors are in series with the line. These faults influence the reliability of the power grid. The causes of these faults are joint failures, circuit break failures and melting of fuse in one or more phases. Figure 6.2 shows the types of open circuit faults in the power systems. For instance, consider a single phase open circuit fault in the power transmission line that causes the increase in the acceleration of the alternator than the actual rated loading. Hence, the alternator runs slightly higher than the synchronous speed. This results in running the alternator runs slightly higher than the synchronous speed. These fluctuations in synchronous speed cause over voltage in grid transmission line.

6.2.2 Short Circuit Faults

The short circuit fault is defined as an abnormal effect due to the low impedance between two points with different potential in the grid. These faults are one of the primary concern in the power systems because they cause huge damage to the power system equipment as well as instability in the entire power system. Further, power system protection is key to safeguard electrical systems and to protect them from loss of synchronism in the synchronous devices such as transformers, generators, and motors.

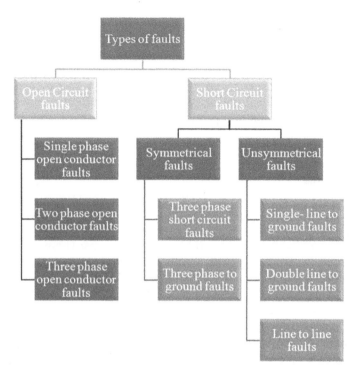

FIGURE 6.1
Types of faults in electrical power grids.

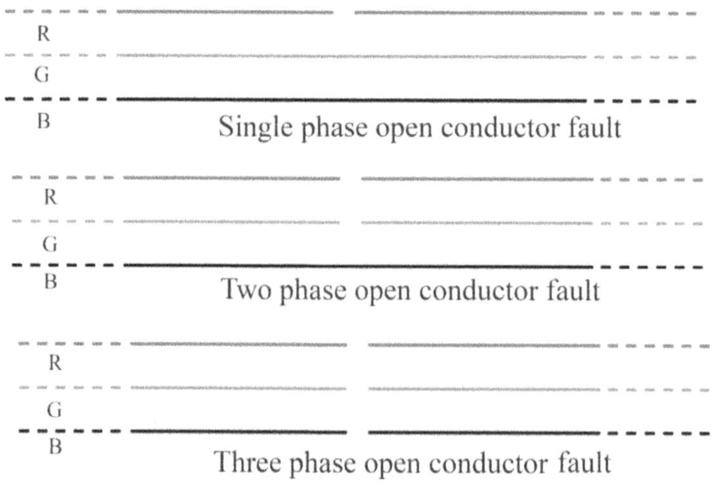

R

G

B   Single phase open conductor fault

R

G

B   Two phase open conductor fault

R

G

B   Three phase open conductor fault

FIGURE 6.2
Types of open circuit faults in the power gird.

6.2.2.1 Symmetric Faults

Symmetric faults are termed "balanced faults" and involve three phases during the fault operation. In electrical power systems these faults are severe and occur rarely. Further, these faults majorly occur at the generator terminals and are classified into the three phase fault (L-L-L) and the three phase to ground fault (L-L-L-G). In an electrical power system the symmetrical faults range from 2% to 5%. However, severity in the faults causes a huge damage to the power devices in the power system. Figure 6.3 shows the types of symmetrical faults in the electrical power systems.

6.2.2.2 Unsymmetrical Faults

Unsymmetrical faults are termed "unbalanced faults". These faults are less severe and very frequent in electrical power systems. Unsymmetrical faults cause uneven fault currents having differing magnitudes and imbalanced phase displacements. These faults are majorly classified into single line to ground (L-G) faults, double line to ground (L-L-G) faults and line to line (L-L) faults as shown in Figure 6.4. Single line to ground (L-G) faults are very frequent and less severe, and range from 70% to 80%. These faults occur when a single conductor makes contact with the neutral conductor or falls to the ground. Double line to ground (L-L-G) faults occur when two conductors make contact with each other as well as with the ground. These faults are very severe and occur in a range of 10% in the electrical power systems. Line to line (L-L) faults occur in the case of a short circuit between the two conductors. These faults are majorly due to heavy winds and the two conductors may swing and come in contact with each other. These faults range from 15% to 20% in the electrical power systems.

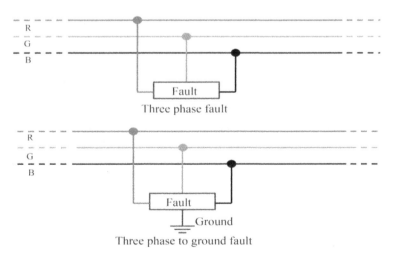

FIGURE 6.3
Types of symmetrical faults in electrical power systems.

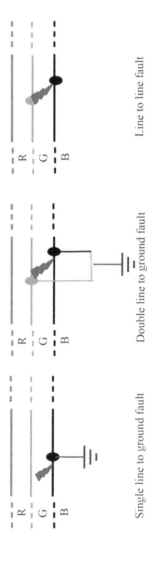

FIGURE 6.4
Types of unsymmetrical faults in electrical power systems.

6.3 Introduction to Superconducting Fault Current Limiters (SFCLs)

The necessity of upgraded distribution systems increases with the increase in the power demand. The increase in the capacity of power generation exceeds the critical current handling capability of the conductor and leads to short circuits, thereby increasing the fault current limit.

Conventional fault current limiters, such as circuit breakers, switch gears, electromagnetic relays and fuses, are not sensitive in detecting the short circuit; external triggering is required and is expensive in electrical component replacement after fault. These are not capable to compensate for the fault current without power interruption. An **SFCL** is an innovative electrical power device which is capable of handling the fault current within the half cycle with improved transient stability as well as reduced current stresses and voltage sags efficiently. Figure 6.5 shows the design of the SFCL. The superconducting tapes are wound over a former (generally, a hollow cylindrical shell made up of PTFE) to replicate a non-inductive coil mandrel. This mandrel is immersed in the Liquid Nitrogen (LN$_2$) bath to cool the superconducting tapes to retain the superconductivity of the superconductor after fault compensation. This SFCL can be integrated in different locations of the grid by conducting the transient stability analysis.

6.3.1 Working of an SFCL

During normal operation, the SFCL works at zero impedance and the large impedance during the fault compensation. The SFCL uses non-linear characteristics and the diamagnetism of the superconductor, when cooled below its critical temperature, depends on the type of SFCL for compensating

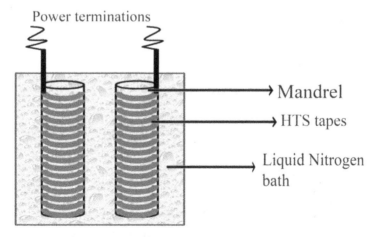

FIGURE 6.5
Superconducting Fault Current Limiter.

the fault current. The V-I characteristics of the normal conductor and the superconductor are shown in Figure 6.6. The normal conductor follows the proportionality law which means that if current increases, then the voltage in the conductor increases as well. However, thermal losses in the normal conductors are huge. The superconductor follows the power law during the short circuit and compensates the fault within milli-seconds. When the fault occurs, if the transport current increases to more than the critical current of the superconductor, the transition from the superconducting state to the normal state takes place due to quenching of the superconductor. At this state, the impedance in the line increases and the fault will be limited within milli-seconds. For instance, the electrical networks in India operate at 50 Hz and the time taken for a single cycle is 0.02 seconds. The selection of the SFCL depends on the fault compensation time and it should be less than 0.02 seconds. Parameters such as activation time, fault compensation time and recovery time are considered while selecting the SFCL for continuous supply of electrical power without interruption.

The SFCL works on the non-linear V-I characteristics of the superconductor

$$V \approx \left(\frac{I}{I_c} \right)^n$$

where, I_c is critical current $I_c = 1$ μV/cm and n is power law index, for superconductor n ranges from 5 to 30.

6.3.2 Characteristics of Ideal Fault Current Limiter

In power transmission and distribution systems, SFCLs revolutionized the way of power system devices to safe guard the electrical power grid with improved performance, efficiency and stability while compensating fault currents. Further, SFCLs have the capability of fast transition from low impedance to high impedance during the fault and automatic recovery to its superconducting state after fault clearance within milli-seconds. The ideal characteristics of SFCLs are listed as follows:

1) Virtually zero voltage drop during rated operation in the smart power systems.
2) Desired fault current limiting capability before the first peak of fault.
3) Multiple fault reduction ability within a short period of time with repeated operations.
4) Automatic and quick recovery to superconducting state after fault compensation.
5) Low cost of maintenance, compact in size, light weight and portable.
6) During normal operation, no negative impacts on the relay and circuit breakers.

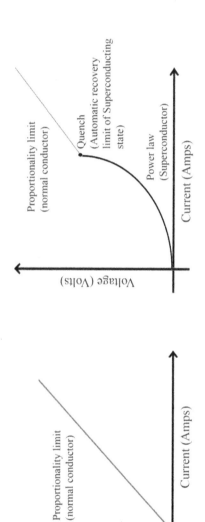

FIGURE 6.6
Voltage vs current characteristics of SFCL.

7) Capable of handling high fault currents without distracting the voltages and angle of stability in modern smart power grids.

8) Large endurance limit for high temperature rise, reliability and durability

However, in practical conditions, definite compromises and trade-offs should be made in different SFCLs to satisfy the abovementioned features.

6.4 Types of Superconducting Materials Used in Fault Current Limiters

In superconductors, the behavior of non-linearity paved the way of developing the superconducting fault current limiters for compensating the fault current to protect the electrical power devices connected to the electrical networks. Superconductors are classified into low temperature superconductors (LTSs) and high temperature superconductors (HTSs). LTSs are cooled using Liquid Helium of critical temperature 4 K and HTSs are cooled using Liquid Nitrogen of critical temperature 77 K. These superconductors are available in the markets as wires, filaments and tapes. In recent decades, ceramic conductors are most significantly used in the construction of an SFCL and the majorly used coated conductors are Yttrium-Barium-Copper-Oxide (YBCO) manufactured in the form of a thin film and Bismuth-Strontium-Calcium-Copper-Oxide (BSCCO). Further, Magnesium Diboride (MgB_2) is another alternative material used in the manufacture of SFCLs due to its low cost. The development in the commercialization of YBCO coated conductors with improved performance and significant drop in the price gained major interest for developing the SFCL. Table 6.1 shows the types of superconductors used in SFCL projects around the world. The commercial 1 G and 2 G tapes used for designing the SFCL is shown in Table 6.2.

6.5 Types of SFCLs

SFCLs are majorly classified into two types

1) Resistive type SFCL (RSFCL)
2) Inductive type SFCL
    a) Shielded Core or Inductive type SFCL
    b) Saturated Core Type SFCL

TABLE 6.1

Types of Superconductors Used in the SFCL Projects Around the World

Type of Superconductor	Number of phases used	Type of SFCL	Power Rating	Year	Company	Country
BSSCCO 2212 Bulk	1 Phase	Resistive	800 A, 8 kV	2001	ABB	Switzerland
		Resistive- Inductive	1.8 kV, 63.5 kV	2008	Nexans	Germany
		Diode Bridge	1.2 kA, 7.2 kV	2002	General Atomics	USA
	3 Phase	Resistive	600 A, 6.9 kV	2004	ACCEL &Nexans SC	Germany
			630 A, 13.2 kV	2007	KEPRI	Korea
		Diode Bridge	200 A, 3.8 kV	2004	Yonsei University	Korea
			1.5 kA, 6 kV	2005	CAS	China
		Iron core DC Biased	1.6 kA, 20 kV	2007	Innopower	China
YBCO thin film	1 Phase	Resistive	40 A, 1 kV	2004	CRIEPI	Japan
			1 kA, 200 V	2004	Mitsubishi	Japan
	3 Phase	Resistive	200 A, 3.8 kV	2004	KEPRI	Korea
			100 A, 4.2 kV	2000	Siemens	Germany
YBCO Coated conductors	1 Phase	Resistive	1.4 kA, 100 V	2001	Alcatel	France
	3 Phase	Resistive	80 kV	2009	IGC Superpower	US
MgB$_2$	–	Resistive	400 A, 6.6 kA	–	Rolls Royce	UK
Non Superconductor	3 Phase	Power Electronics	1.2 kA, 8 kV	2004	Powell	US
			25MVA, 6.9 kV	2004	Siemens	Germany

TABLE 6.2

Commercial 1 G and 2 G Tapes Used for Designing the SFCL

Manufacturer		AMSC (344 S)	SuperPower (SF12050)	SuperPower (SF12100)	Sumitomo (Di-BSCCO)
Category of Superconductor		2G (coated conductor)	2G (coated conductor)	2G (coated conductor)	1G (Multifilament)
Type of Superconductor	Material	YBCO	YBCO	YBCO	BSCCO-2223
	Thickness (μm)	~1	~1	~1	< 1
Substrate	Material	Hastelloy (Ni-5% at W)	C- 276®Hastelloy (Ni-W-Cr)	C- 276®Hastelloy (Ni-W-Cr)	NA
	Thickness (μm)	75	50	100	NA
Stabilizer	Material	Stainless steel (SUS 316L)	Stabilizer free	Stabilizer free	AgAuMg matrix
	Thickness (μm)	120	NA	NA	~0.2
Buffers		Y2O3/YSZ/CeO2	Al2O3/Y2O3/MgO/LMO	Al2O3/Y2O3/MgO/LMO	--
Process		RABiTS/MOD	IBAD	IBAD	--
Tape Dimensions	Width (mm)	4.30	12	12	4.5
	Thickness (mm)	0.218	0.055	0.105	0.31
Critical current, I_c@77K, self-filed (A)		106	268	375	170
Critical current density, J_c (A/mm²)		136	210	290	--
Critical Temperature		90	90	90	110

6.5.1 Resistive Type SFCL (RSFCL)

A resistive SFCL is simple and compact compared to the inductive SFCL and is connected in series with the smart power grid. It constitutes a superconducting element for triggering the fault current using the natural characteristic of non-linearity in the superconducting material. During normal operation, the RSFCL remains in the superconducting state and the resistance offered by the material is zero. When the fault occurs, the transport current carried by the superconductor increases and if that current reaches the critical current of the superconducting material, the superconductor switches from the superconducting state to the normal state within milli-seconds. Therefore, the increase in the resistance of the superconductor limits the fault current. A resistor or an inductive shunt is positioned in parallel to the superconducting element as shown in Figure 6.7 to avoid hotspots during the quench, to regulate the limiting current and to sustain from over-voltages during fault compensation. A cryogenic coolant such as Liquid nitrogen is circulated to retain the superconductivity of the RSFCL. The ohmic losses during the fault current are dissipated to the cryogenic coolant and the recovery from the normal state to the superconducting state in the superconductor takes place automatically. "Recovery Time" is defined as the rate at which the RSFCL recovers from the normal state to the superconducting state, during quench. It is one of the critical parameters used while designing the SFCL in the smart grids. During fault compensation, the boiling of the cryogenic coolant occurs and the superconductor remains in the superconducting state till the entire coolant evaporates. In such cases, at an instance, the RSFCL losses its superconductivity and leads to the potential failure with hot spot formation called "Run-off".

6.5.1.1 Advantages of RSFCLs

1) The design of RSFCLs is compact, simple and light in weight.
2) Automatic triggering during the fault current.
3) Fault compensation is fast and effective when the fault current in the line exceeds the critical current of the superconductor.
4) Fail safe during fault limitation.
5) Good operational benefits.

6.5.1.2 Disadvantages of RSFCLs

1) The requirement of the superconductor is higher while designing the RSFCL, which is directly associated with the required quench resistance and the rise in temperature of the superconductor during the fault.
2) Formation of hotspots during fault compensation. This is due to the quench at some point on the superconductor and it is not possible

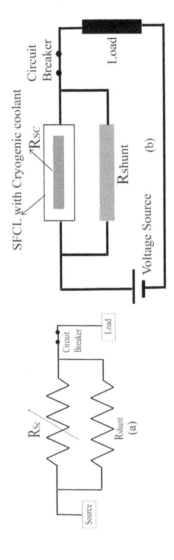

FIGURE 6.7
(a) Line diagram and (b) electrical circuit diagram of RSFCL.

to guarantee quenching at every point on the superconductor at the same time.

3) High energy losses during fault limitation. When the RSFCL encounters the fault current, quenching of the SFCL coil begins and becomes resistive to compensate the level of fault current. During fault operation, high energy is dissipated by the superconductor due to Joule heating losses before fault compensation.

4) Long recovery time from the normal state to the superconducting state due to Joule heating losses results in the rise in temperature of the superconductor. Recovery time ranges from several seconds to several minutes after fault compensation.

5) There are few technical challenges while integrating the SFCL to external power systems such as maintaining the room temperature and cryogenic interface, as heat in the leaks from the current leads may cause insulation problems and space requirements to handle higher power densities in the cryogenic environment.

6.5.2 Inductive Type SFCL

The inductive type SFCL has the simplest design similar to the transformer connected in series to the smart grid. This transformer constitutes a conventional primary coil and a closed superconducting ring as the secondary coil. During the normal operation, the superconducting rings are de-exited and the impedance is due to the primary winding resistance and leakage in inductance. Hence, the impedance in the SFCL is low and approximately equal to the leakage reactance. During the fault operation, the transport current increases over the critical current of the superconductor ring and the transformation from the superconducting state to the normal state takes place. In such a case, the SFCL encounters a high impendence, approximately equal to the main field reactance.

6.5.2.1 Shielded Core SFCL

The shielded core SFCL works on the principle of Meissner effect, where the superconductor acts as a perfect diamagnetic conductor because in the superconducting state it will expel the applied magnetic field. This SFCL works similar to the transformer; the primary iron coil is connected to the power line and is short circuited with the single turn secondary coil which is made of a cylindrical superconducting element. During normal operation, low impedance is induced in the line and the superconductor effectively monitors the primary iron core. When the fault occurs, the transport current and the magnetic field increase higher than the critical level. Therefore, the flux penetrates into the primary iron core because the superconductor is unable to shield and high impedance is introduced in the line of the smart grid that is to be protected.

Figure 6.4 shows the circuit diagram of shielded core SFCL; the iron core primary windings are connected to the main line of the grid and are out of the cryogenic environment. During normal operation, the magnetic flux generated by the primary windings is expelled from the secondary core of the superconductor and does not penetrate into the iron core. Hence, the magnetization losses are zero in the iron core and low impedance is available in the grid. The secondary windings of the shielded SFCL consist of superconductive windings and a normal bypass (Resistance). During fault, the superconductor switches from the superconducting state to the normal state and the flux pinning through the superconductor takes place. Since, YBCO-based superconductors are used more predominantly in the shielded SFCL, the change in the state of superconductor dissipates huge amounts of thermal energy. The normal bypass coil allows the current flow in this state. The superconductor expels the magnetic flux till the fault current is lesser than the critical current and if it exceeds the run-off of superconductor takes place. The voltage induced in the secondary bypass due to coupling with the iron core counters the fault current to compensate the current in the primary coil (Figure 6.8).

6.5.2.2 Saturated Core SFCL

The increase in the impedance in RSFCL and shielded core SFCL is due to the quenching of coated conductors during the fault. However, the dynamic behavior of the magnetic iron core in the saturated core FCL increases the impedance in the AC Line. The Saturated core FCL consists of two iron cores, AC windings, DC control unit, DC power supply, and superconducting DC windings. The two iron cores that are utilized for each half of the cycle and the normal conductors are wound around the cores of AC windings to induce impedance in series with the line of the grid. The two iron cores are saturated using the DC magnetic field generated by the superconducting coil which is wrapped around each core to maintain the constant current that provides magnetic bias. During normal operation, the saturated core SFCL works analogous to the air core reactor where the relative permeability is unity and the iron cores are fully saturated by the superconducting coil. When the fault occurs, the surge in the transport current takes place and instantly the monitoring system in the DC control unit switches the existing DC current using the power electronic switches within a few milli-seconds. For instance, the integrated gate commutated thyristor (IGCT) or the insulated gate bipolar transistor (IGBT) is used in the DC control unit. During the fault operation, the impedance in the power line will increase in each half cycle and the saturation across the cores is driven out due to induced EMF in the alternative peaks. The advantage with the saturated core FCL is that there is no necessity of transformation from the superconducting state to the normal state during fault compensation. However, for designing this FCL requires huge amounts of iron since it consists of two iron cores. The saturated core SFCL

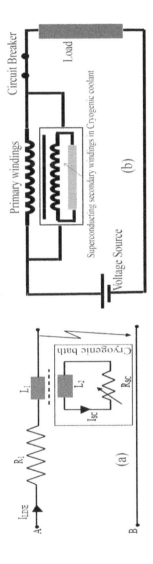

FIGURE 6.8

(a) Line diagram and (b) circuit diagram of shielded inductive type SFCL.

can be made compact by replacing the coated conductors with HTS DC field which is effective in reducing operating losses (Figure 6.9).

6.5.3 Advantages of Inductive SFCLs

1) Intrinsically fail safe. When the DC bias fails to supply the power, the iron core gets de-saturated and a high impedance will be produced from the primary windings to compensate the fault current.

2) The recovery of the superconductor after fault compensation is faster because during the fault the superconducting state is retained in the coil.

3) High voltage power transmission design is also possible because there is no necessity of any interface between room temperature and the cryogenic environment in the line of the power grid.

4) There are no thermal losses through the current leads. Hence, the power for circulating the cryogenic coolant may be lower.

5) Hotspots in the superconductor can be easily reduced by optimizing the turns ratio.

6.5.4 Disadvantages of Inductive SFCLs

1) Huge in construction and heavier in weight compared to RSFCLs due to the availability of iron cores while designing the inductive SFCL.

2) During normal operation, the SFCL encounters significant losses in the iron cores and primary windings.

3) Complexity with supply of current to the superconducting windings. For instance, if any unwanted current is introduced into the DC bias coil, it results in high voltage across the superconducting coil which could probably damage the supply of DC bias.

6.6 Other Types of SFCLs

6.6.1 Resistive Magnetic SFCL

The resistive magnetic SFCL employs parallel inductance to reduce the hotspot in the superconductors during fault operation. This SFCL consists of a normal conducting coil which is coaxial and exterior to the superconducting tube. During normal operation, the normal property of superconductors is used, that is, the magnetic flux lines are expelled due to the surface currents until the point of critical current of superconductor is

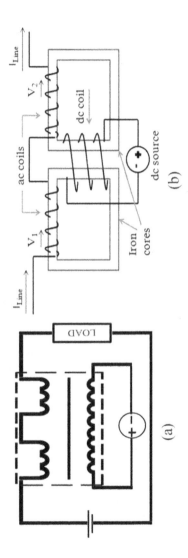

FIGURE 6.9
(a) Line diagram and (b) electrical circuit diagram of saturated core SFCL.

reached. When the fault occurs, the quench of the superconductor starts from the weakest point and the increase in the resistance of the super-conductor allows the current to flow through the bypass parallel coil. The critical current in the superconductor is lowered by the magnetic field of the coil and the quench is accelerated to mitigate the hotspot formation in the superconductor.

6.6.2 Bridge SFCL

In the bridge SFCL, solid state technology is employed to control the surge of current through a superconducting inductance. In this SFCL, the fault current is not compensated by quenching the superconductor by transition from the superconducting state to the normal state during fault. Instead, the superconducting material is used in the DC state of unconstrained transport characteristics. It consists of a normal conducting material and uses the principle of inductive impedance for limiting the fault current. The construction of the bridge SFCL involves four diodes, DC bias supply and the superconducting coil (L) as shown in Figure 6.10. During normal operation, the amplitude in the superconducting coil is larger than the DC circuit. However, during the fault, when the amplitude of the current increases, the half cycle in the diode bridges D3 and D4 resist the current flow and the other negative half diode bridge D1 and D2 resist the current. The superconducting coil which is in the line automatically compensates the fault with a large impedance L. The requirement of the DC bias supply and the power diode bridge made this SFCL complex. When the electrical power system is operated at normal conditions, there is no necessity of using the superconducting material because only DC current is supplied which makes the use of superconductors an ideal choice. The requirement of the superconducting material

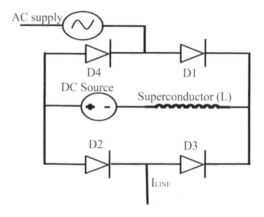

FIGURE 6.10
Line diagram of Bridge SFCL.

is to minimize transmission losses. Further, the diodes can be replaced by Thyristors for ease in the operation of this SFCL after fault compensation.

6.6.2.1 Advantages of Bridge SFCL

1) AC losses in the superconducting coil are zero, since DC current is used in operating the superconducting coil.
2) Fast recovery because the superconducting coil remains in the superconducting state during fault compensation.
3) The level of fault triggering can be adjusted by the DC source.
4) There is no necessity of any interface between room temperature and the cryogenic environment in the line of the power grid.

6.6.2.2 Disadvantages of Bridge SFCL

1) Semiconductors are used in this SFCL. Hence, significant AC losses are encountered during operation.
2) Intensively fail safe operation is not possible.
3) During the fault operation if one of the semiconductors fails, there is a chance of short circuit and fault compensation is not possible.

6.6.3 DC Biased Iron Core SFCL

The DC biased iron core SFCL consists of two iron cores positioned in series of the fault current path. During normal operation, these coils are saturated using DC bias current and the inductance in the line is low. During fault operation, these coils are expelled from the saturation mode results in the rapid increase in coil inductance and fault compensation takes place. The advantage of this SFCL is less volume of superconducting material and cryogenic coolant is required, since iron cores are used in designing such SFCLs. Hence, this SFCL will be heavier than other SFCLs.

6.6.4 Solid State FCL (SSFCL)

The solid state FCLs are developed to overcome limitations such as slow recovery after fault compensation and inefficient performance during fault operation. In this FCL, semiconducting devices such as GTO, IGBTs and ETO are used. The conventional design of circuit breakers is modified with the solid state switches. These FCLs offer robustness, efficient operation during fault and significant functionality and are positioned in the potential path of the current flow and switch off during the fault operation. The disadvantage of these FCLs is switching losses due to harmonics in the semiconductors. The theoretical advantages of these FCLs are compact size, cryogenic environment free, instant recovery after fault compensation, minimum losses, modularity and expandability.

6.6.5 Fault Current Controllable SFCL

The power electronic semiconducting devices such as Integrated Gate-Commutated Thyristors (IGCTs) and Insulated Gate Bipolar Transistors (IGBTs) have the capability to compensate the fault current by turning off the devices and limiting the impedance. The advantage of this SFCL is that adjustable triggering during the fault current compensation is possible. Hence, a complete control over fault compensation makes them fault current controllers (FCCs). This SFCL exhibits similar characteristics to that of the diode bridge SFCL explained in Section 6.4.3.2. However, the semi-conducting devices required are less compared to the diode bridge SFCL. Normal conductors can also be used for the operation of this SFCL. These will increase the thermal losses and size of the SFCL.

Various fault current limiters are compared depending on losses, type of triggering, recovery, size or weight and distortion as shown in Table 6.3.

6.7 Applications of SFCL in Power Transmission and Distribution Systems

The probability of short circuits in the power grids is increasing due to growth in inter-connections of electrical power systems. The potential advantage of the SFCL provides solution for the protection of the electrical power grid. However, the transient electromagnetic behavior under short circuit is to be investigated for installing the SFCL at different locations in the distribution grid. The transient recovery voltage (TVR) and the transient overvoltage in the grid are improved by installing the SFCL during fault operation. Further, installation of the SFCL reduces the voltage sag in the grid after fault compensation. Hence, the SFCL can be integrated at different locations in power grid at the distribution level and at the transmission level. Many studies show the advantages in the integration of the SFCL in electrical power systems for reduction in current stresses in the power devices, enhancement in transient stability during the fault and voltage sag and dip reduction. Figure 6.11 shows the integration of the SFCL at different locations in the power grid such as Coupling at Feeder Transformer and Busbar, at local power generation and at feeder generator. For instance, an SFCL is connected at a local feeder location and during the fault huge currents are passed through the SFCL. When the fault current exceeds the critical current in the superconductor the fault will be compensated in the first half cycle.

TABLE 6.3

Comparison of Types of Fault Current Limiters

Type of fault Triggering	Type of FCL	Losses	Cooling system	Maximum rating	Activation time	Recovery time	Status
Passive type	Resistive SFCL (RSFCL)	Hysteretic	Yes	138 kV, 0.9 kA	Less than 1/4 cycle	msto 2 s	Designed and tested
	Inductive SFCL	Hysteretic, Joules heating and saturation state	Yes	11 kV, 2 kA	Immediate	< 5 ms	R&D stage
Active type	Hybrid Resistive SFCL	Hysteretic	Yes	12 kV, 2 kA	100 ms	Controlled	Research stage
	Solid state SFCL (Circuit Breakers)	Hysteretic	No	69 kV, 3 KA	Less than 10 µs	Controlled	Development stage

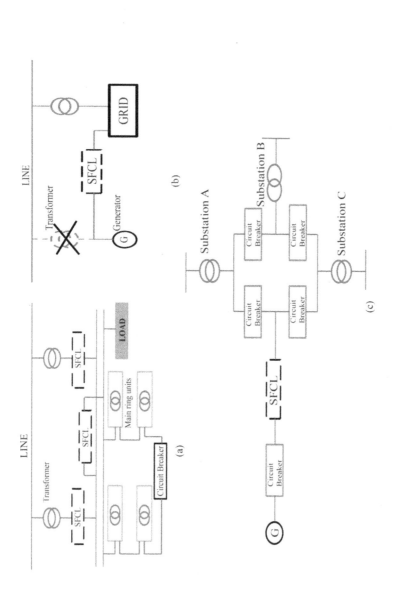

FIGURE 6.11

Integration of SFCL in different locations: (a) coupling at feeder transformer and busbar, (b) local power generation, and (c) at feeder generator.

The transition from the superconducting state to the normal state takes place during fault compensation and the maximum short circuit is limited to protect the grid (Table 6.4).

6.8 Thermo-Electrical Design of SFCL

6.8.1 Resistive Superconducting Fault Current Limiter (RSFCL)

For designing an RSFCL, the transient stability analysis during the fault has to be estimated using the electro-thermal analysis using the nominal current–dependent Time-Current characteristics [1–4].

6.8.1.1 Basic Parameters

The basic parameters include the selection of ceramic-coated superconducting material such as YBCO or BSSCO. For instance, depending on the material select the tape and specify material properties such as critical current density (J_C), critical temperature (T_∞), resistivity (ρ), conductor index number (n), specific heat (C_v) and minimum critical current at the operating temperature, and tape dimensions such as width of conductor (w) and thickness of conductor (t). Further select the cryogenic coolant to be used for cooling the HTS tapes to retain the superconductivity during fault and specify the

TABLE 6.4

Locations of SFCL Installation With Power Ratings and Type of SFCL Used

Location of SFCL installation	Rated Voltage	Rated current	Type of SFCL	Country	Year
Outgoing feeder	9 kV	1 Ka	RSFCL	Italy	2016
		220 A			2012
	12 kV	880 A		Germany	2011
	15 kV	1.25 kA	Inductive	US	2009
	35 kV	1.5 kA		China	2008
Bus-bar Coupling	12 kV	1.6 kA	RSFCL	UK	2015
		1.05 kA			
		400 A			2012
		100 A			2009
Coupling the local generation	10 kV	815 A	RSFCL	Germany	2016
HTS cable protection	12 kV	2.3 kA	RSCL	Germany	2014
Local grid protection and Neutral ground Resistor (NGR)	13.8 kV	1 kA	RSFCL	US	2013
Secondary side of transformer	115 kV	500 A	RSFCL	Thailand	2016
Network Coupling	220 kV	800 A	Inductive	China	2012

cryogenic bath temperature (T_a). For electro-thermal analysis, the maximum fault limit time [4] should be fixed and it should be less than the 0.02 seconds as per the Indian scenario. For example, the fault limit time ranges from 80 ms to 120 ms [5]. The fault behavior during the analysis can be studied using the J-B-T curve of the superconductor [6]. The critical temperature of the superconductor depends on the J (T) and B (T). The basic parameters to be considered while designing the SFCL are: [7]

1) **Transport Current**

    The manufacturer specifies the peak current depending on the tape and the RMS current should be calculated to predict the maximum possible current that can be transported through the tape.

$$I_p = \sqrt{2} I_{rms} \tag{6.1}$$

2) **Area of the Superconductor**

    The current carrying capacity depends on the dimensional specifications of the tape such as width and thickness [8].

$$A_{HTS} = w * t \tag{6.2}$$

Area of superconducting tape in terms of critical current and critical current density is shown in Equation (6.3)

$$A = \frac{I_c}{J_c(T)} \tag{6.3}$$

3) **Length of the Tape**

    Electrical field in a HTS tape can be calculated as follows

$$E = \frac{dv}{dl} \tag{6.4}$$

where, v is transmission voltage

    Length of the superconducting tape required for designing the SFCL should be estimated.

$$L_{HTS} = \frac{\sqrt{2} V_{rms}}{E_{peak}} \tag{6.5}$$

Length of the superconducting tape can also calculated using Equation (6.6) from the parameters such as first peak current (I_{rp}) and SFCL resistance as follows [9].

$$L_{HTS} = \frac{R_{SFCL}}{E_C} * \frac{I_C^{\,n}}{I_{rp}^{\,n-1}} \tag{6.6}$$

4) **Number of HTS Tape Windings Over Mandrel**

The HTS tapes are wound over the former (a hollow cylindrical shell) of the mandrel in clockwise and anticlockwise directions depending on the inductance of the coil. The total number of turns also depends on the diameter of the mandrel and the dimension of the HTS tapes and the direction in which the HTS tapes are wound while designing the SFCL. Figure 6.12 shows the thermo-electrical strategy for designing the SFCL for compensating the fault current.

6.8.2 Electrical Strategy

The normal characteristics of superconductors such as zero resistivity and perfect diamagnetism when cooled below to the critical temperature are used for designing the SFCL. The designed SFCL is placed in series to the line and a shunt resistance is connected in parallel to the superconductor. When the fault occurs, the transport current exceeds the critical current (I_C) and the transition from the superconducting state to the normal state takes place that results in quench; thereby, the resistance increases in the superconductor. However, during the quench hotspots are formed in the superconductor because of non-uniformity in material and evaporation of the cryogenic coolant while cooling of the superconductors [10–13]. In order to reduce the

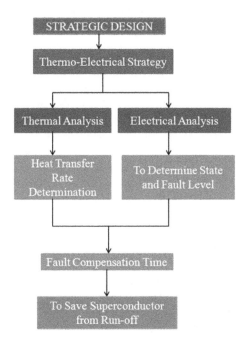

FIGURE 6.12
Thermo electrical strategy of SFCL.

formation of the hotspots a shunt is connected to the superconductor. The current flows through the shunt resistance when the SFCL is quenched and the fault will be compensated within milli-seconds.

Figure 6.13 shows the transition of the HTS tape from the superconducting state to the normal state during the fault in the electrical power systems. The E-J characteristic curve [14–17] is used to discuss the transition of the super-conductor. They are classified into the Flux creep state where the supercon-ductor has the properties of zero resistivity and diamagnetism, Flux flow state where the pinning of flux lines through the superconductor takes place and Normal state where the proportionality law of the superconductor occurs.

The E-J characteristic curve of the superconductor can be calculated using the E-J power law as shown in Equation (6.7) [18–20] and the E-I characteris-tic of the superconductor can be estimated using Equation (6.8) [2, 21]

$$E = E_0 \left(\frac{J}{J_C} \right)^n \tag{6.7}$$

$$E = E_0 \left(\frac{I}{I_C} \right)^n \tag{6.8}$$

where, n is power index term, I is operating current in the power system, I_C is critical current and E_0 is electrical field of the superconductor. The calculation of the E-J characteristic is important to estimate the fault compensation time.

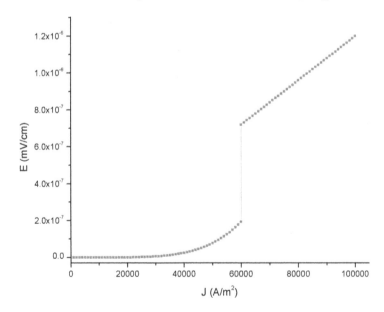

FIGURE 6.13
E-J characteristic of superconductors with three states of transition.

6.8.2.1 State-1 (Superconducting State) ($\rho=0$)

The state at which diamagnetism and zero resistivity are exhibited in the superconductor when cooled below its critical temperature and follows the power law. The E-J characteristic of the superconductor state is shown in Equation (6.9).

$$E^{(1)}(J,T) = E_0 \left(\frac{J}{J_C}\right)^{\alpha(T)} \tag{6.9}$$

The power law index term $\alpha(T)$ is considered as the maximum of index term in the superconducting state β and the operating temperature index term $\alpha'(T)$.

$$\alpha(T) = \max\left[\beta, \alpha'(T)\right]$$

where,

$$\alpha'(T) = \frac{\log\left(\dfrac{E_0}{E_C}\right)}{\log\left(\dfrac{J_C(T_a)}{J_C(T)}\right)^{\left(1-\frac{1}{\beta}\right)}\left(\dfrac{E_0}{E_C}\right)^{\frac{1}{\alpha(T_a)}}} \tag{6.10}$$

6.8.2.2 State-2 (Flux Flow State) ($\rho=\rho$ (J))

The state at which the flux lines start pinning through the superconductor and resistivity increases which can be studied using the J-B-T curve. The E-J characteristic equation in this mixed state is given by Equation (6.11).

$$E^{(2)}(J,T) = E_0 \left(\frac{E_C}{E_0}\right)^{\frac{\beta}{\alpha(T_a)}}\left(\frac{J_C(T_a)}{J_C(T)}\right)\left(\frac{J}{J_C(T_a)}\right)^{\beta} \tag{6.11}$$

6.8.2.3 State-3 (Normal State) ($\rho=constant$)

The state at which the superconductor acts as a normal conductor which completely allows magnetic flux lines to flow through it, and the resistivity of the conductor increases drastically compared with other two states. In this state it follows the proportionality law because of the ohmic behavior. The E-J characteristic in normal state of the conductor is given in Equation (6.12)

$$E^{(3)}(J,T) = \rho(T_C)\frac{T}{T_C}J \tag{6.12}$$

where, E = 1e-6 V/m, $1000 \leq J_C \leq 10000$ A/cm^2, $5 \leq \alpha \leq 15$, $0.1 \leq E_0 \leq 10$ mV/cm, $2 \leq \beta \leq 4$, $100 \leq \rho \leq 2000$ μΩ-cm at an operating temperature of 77 K.

From the J-B-T curve the transition from a superconducting state to a normal state is observed due to critical current density, critical magnetic field and critical temperature. Hence the effect of critical temperature (T_C) on the transition behavior of superconductor with respect to time is shown in Equation (6.13) [22]. The necessity of evaluating the temperature-dependent current density is to calculate the run-off of the superconductor and the volumetric rate of LN_2 required for cooling the SFCL.

$$E(T,t) = \begin{cases} E_c \left(\dfrac{J(t)}{J_c(T(t))} \right)^{\alpha} \\ \quad \text{for } E(T,t) < E_0 \text{ and } T(t) < T_c \\ \\ E_0 \left(\dfrac{E_c}{E_0} \right)^{\frac{\beta}{\alpha}} \left(\dfrac{J_c(T_a)}{J_c(T(t))} \right) \left(\dfrac{J(t)}{J_c(T_a)} \right)^{\beta} \\ \quad \text{for } E(T,t) > E_0 \text{ and } T(t) < T \\ \\ \rho(T_c) \dfrac{T(t)}{T_c} J(t) \\ \quad \text{for } T(t) > T_c \end{cases} \tag{6.13}$$

The flowchart used for analyzing the electro-thermal behavior of R-SFCL is shown in Figure 6.14. The basic parameters which are discussed in Section 6.1.1 at the cryogenic bath temperature should be initialized and the maximum compensation time during fault should be less than 0.02 seconds for the Indian scenario. During the fault operation, if the operating temperature and electrical filed in the superconductor are less than the critical parameters then the superconductor exists at flux creep state. If the operating temperature is less than the critical temperature and the electrical field is higher than the critical electrical field the flux flow state of the superconductor exists. If the operating temperature and the electrical field are higher than the critical parameters, the normal conducting state exists. The time taken for the fault compensation should be calculated and if the total time taken for the fault compensation is greater than or equal to the maximum considered fault time the parameters are considered as the final parameters. The fault compensation time during the fault current can be estimated using electrical strategy.

6.8.3 Thermal Strategy

The SFCL immersed in the cryogenic bath with LN_2 as a cryogenic coolant starts evaporating due to the increase in the thermal losses during the transition from the superconducting state to the normal state and the resistance offered by the superconductor in the normal state. Hence, the rise in the temperature of cryogenic bath leads to run-off of the superconductor due to hot

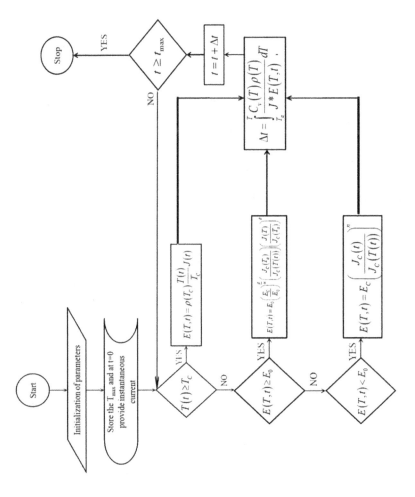

FIGURE 6.14
Flow chart for simulating the strategic design of R-SFCL.

spot formation. Therefore, there is a necessity to calculate the heat transfer rate in the SFCL to retain the superconductivity of the HTS tapes and the transition from the normal state to the superconducting state after fault compensation. Hence, for the protection of electrical power systems as well as for efficient power transmission and distribution, cooling of the coated conductor after fault compensation is essential to regain its superconductivity. The heat transfer rate through the HTS tapes to the cryogenic coolant bath can be estimated as follows.

Heat Capacity of the superconductor (J/K)

$$C_{SC} = C_v * V_{SC} \qquad (6.14)$$

where,

C_v is volumetric specific heat of superconductor (J/Km³) at cryogenic environment temperature

V_{SC} is Volume of superconducting tape used

$$V_{SC} = l_{SC} * a_{SC} \qquad (6.15)$$

where, l_{SC} is length of superconducting tape and a_{SC} is area of superconducting tape

Convective thermal resistance between the superconducting tapes and cryogenic bath is given by Equation (6.16)

$$R_{conv} = 1 / h_{sc-cb} * A_{mandral} \qquad (6.16)$$

where, R_{conv} is convective thermal resistance in (K/W), h_{sc-cb} = convective heat transfer coefficient of cryogenic coolant (W/m²K) and $A_{mandral}$ is area of the mandrel

Heat dissipated by the superconductor due to ohmic losses is given by Equation (6.17)

$$Q = i^2 R \qquad (6.17)$$

The rate of heat generation in the superconducting tapes is given by Equation (6.18)

$$Q_{SC}(t) = \int i_{SC}(t)^2 * R_{SC}(t) dt \ J \qquad (6.18)$$

Resistance offered by the superconductor due to transition from the superconducting state to the normal state is given in Equation (6.19)

$$R = \frac{\rho l}{a} \qquad (6.19)$$

where, ρ = Resistivity of the superconductor

$$\rho = \frac{R*a}{l} \tag{6.20}$$

$$\rho = \frac{E}{J} \tag{6.21}$$

Equating Equation (6.19) and Equation (6.20)

$$R = \frac{E*l}{J*a} \tag{6.22}$$

The rate of change of resistance in the superconductor due to fault current [23–27]

$$R_{SC} = \frac{E(t,T)*l_{SC}}{J(t)*a_{SC}} \tag{6.23}$$

$$R_{SC}(t+\Delta t) = \frac{E(t)*l_{SC}}{i_{SC}(t)} \tag{6.24}$$

$$R_{SFCL}(t) = R_m \left(1 - \exp\left(\frac{-t}{T_{SC}} \right) \right) \tag{6.25}$$

where, R_m is maximum Superconductor resistance

The heat transfer rate from the superconducting tapes to the cryogenic cooling bath

$$Q = \frac{\Delta T}{R} = \frac{T(t)-T_a}{R_{conv}} \tag{6.26}$$

Substituting Equation (6.7) in Equation (6.26)

$$Q_{removed} = h_{sc-cb}*(A)_{mandral}*(T(t)-T_a) \tag{6.27}$$

The rate of heat dissipated by the superconducting tapes is determined by Equation (6.28)

$$Q_{removed}(t) = \int h_{sc-cb}*(A)_{mandral}*(T(t)-T_a)dt \tag{6.28}$$

Transient system stability can determined as [2, 14, 15, 28–31]

$$\nabla(k(T)\nabla T) + q(T) = C(T)\frac{\partial T}{\partial t} \tag{6.29}$$

$$k(T)\frac{\partial^2 T}{\partial x^2} + q(T) = C(T)\frac{\partial T}{\partial t} \tag{6.30}$$

$$Q_{generated} - Q_{removed} = \rho C_v \frac{dT}{dt} \tag{6.31}$$

$$i^2 R_{SFCL}(t) - Q_{removed} = \rho C_v \frac{dT}{dt} \tag{6.32}$$

$$E_z J_z - Q_{removed} = \rho C_v \frac{dT}{dt} \tag{6.33}$$

From Equation (6.31) [27, 29, 30]

$$Q_{generated} - Q_{removed} = \rho v_{SC} C_v \frac{dT}{dt} \tag{6.34}$$

$$(Q_{generated} - Q_{removed})dt = \rho C_{SC} dT \tag{6.35}$$

$$T(t) - T_a = \frac{1}{\rho C_{SC}} \int_0^t (Q_{generation} - Q_{removed})dt \tag{6.36}$$

$$T(t) = T_a + \frac{1}{C_{SC}\rho} \int_0^t (Q_{generation} - Q_{removal})dt \tag{6.37}$$

The time taken for limiting the fault current by the SFCL can be calculated as follows

$$\frac{dT}{dt} = \frac{1}{C_{SC}}(Q_{generated} - Q_{removed}(t)) \tag{6.38}$$

$$Q_{generated} = i^2 R_{SC} \tag{6.39}$$

$$Q_{generated} = i*i*R_{SC} \quad \left[\therefore R_{SC} = \frac{\rho_{SC}*l_{SC}}{a_{SC}}, \rho_{SC} = \frac{E(t,T)}{J} \right] \tag{6.40}$$

$$Q_{generated} = i(t)*E(t,T)*l_{SC} \tag{6.41}$$

Substituting Equation (6.41) in Equation (6.38)

$$\frac{dT}{dt} = \frac{1}{C_{SC}}\left[i(t) * E(t,T) * l_{SC} - \frac{T(t) - T(a)}{R_{conv}}\right]$$ (6.42)

$$dt = \frac{dT}{\frac{1}{C_{SC}}\left[i(t) * E(t,T) * l_{SC} - \frac{T(t) - T_a}{R_{conv}}\right]}$$ (6.43)

Taking current as constant, the above equation can be modified as [23]

$$\int dt = \int \frac{dT}{\frac{1}{C_{SC}}\left[IE(T)l_{SC} - \frac{T(t) - T(a)}{R_{conv}}\right]}$$ (6.44)

$$E(t) = E_0\left(\frac{E_C}{E_0}\right)^{\frac{\beta}{\alpha(T_a)}}\left(\frac{J_C(T_a)}{J_C(T)}\right)\left(\frac{J(t)}{J_C(T_a)}\right)^{\beta}$$ (6.45)

$$J_C(T) \approx J_{C(T_a)}\left(\frac{T_C - T(t)}{T_C - T_a}\right)$$ (6.46)

$$E(t,T) \approx \rho\left(\frac{T}{T_C}\right)J(t), \quad T(t) \geq T_C$$ (6.47)

$$E(T) = E_0\left(\frac{E_C}{E_0}\right)^{\frac{\beta}{\alpha(T_a)}}\left(\frac{T_C - T_a}{T_C - T}\right)\left(\frac{J(t)}{J_C(T_a)}\right)^{\beta}$$ (6.48)

$$E(T) = E_0\left(\frac{E_C}{E_0}\right)^{\frac{\beta}{\alpha(T_a)}}\left(\frac{T_C - T_a}{T_C - T}\right)\left(\frac{I}{J_C(T_a)a_{SC}}\right)^{\beta}$$ (6.49)

$$IE(T)l_{SC} = k\frac{T_C - T_a}{T_C - T}$$ (6.50)

$$k = IE_0\left(\frac{E_C}{E_0}\right)^{\frac{\beta}{\alpha(T_a)}}\left(\frac{I}{J_C(T_a)a_{SC}}\right)^{\beta}l_{SC}$$ (6.51)

$$k = E_0\left(\frac{E_C}{E_0}\right)^{\frac{\beta}{\alpha(T_a)}}\left(\frac{I^{\beta+1}}{\left(J_C(T_a)a_{SC}\right)^{\beta}}\right)l_{SC}$$ (6.52)

$$t = C_{SC} \int \frac{dT}{\left[k\left(\dfrac{T_C - T_a}{T_C - T} \right) - \left(\dfrac{T(t) - T_a}{R_{conv}} \right) \right]} \tag{6.53}$$

The E-J characteristic as a function of time is given by [16, 17]

$$E = \sqrt{\frac{C_{veff} * \Delta T * \rho_{eff}}{\Delta t}} \tag{6.54}$$

By solving the above equation [32]

$$\frac{E * J}{C_{v-eff} * \rho_{eff}} = \frac{\Delta T}{\Delta t} \tag{6.55}$$

$$\Delta t = \int_{T_a}^{T} \frac{C_v(T)\rho(T)}{J * E(J,T)} dT \tag{6.56}$$

The time rate change of temperature is given in Equation (6.57) [33]

$$T(t) = \int_{T_a}^{T} \frac{J^2 \rho_{SC}}{C_{SC}} dt \tag{6.57}$$

The thermo-electric analysis of the SFCL during normal and fault operation with and without SFCL is shown in Figure 6.15. The operating current in the line is increased drastically during the fault. The fault is compensated within milli-seconds and in the first cycle after the fault operation recovery of RSFCL begins. With the integration of RSFCL in the electrical power system 60% of the faults are reduced <5 ms. Hence, the installation of RSFCL is one of the better option due to advantages offered by the RSFCL than inductive SFCL.

6.9 Inductive Superconducting Fault Current Limiter

The critical current density in the superconductors is calculated by

$$E = \sqrt{\frac{\rho}{t} \frac{dp}{dA}} \tag{6.58}$$

where, ρ is resistivity of the superconductor, t is time, $\dfrac{dp}{dA}$ is surface power dissipation density

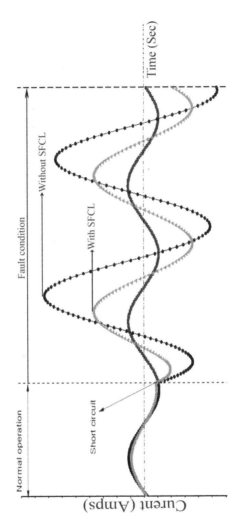

FIGURE 6.15
Operation of SFCL during fault condition.

AC losses in the inductive SFCL can be estimated using the H-formulation, since this SFCL works on the natural behavior of diamagnetism in the superconductors [34].

The governing equations for inductive SFCL can be modeled using the H-formulation

From Faraday's law

$$\nabla \times \mathbf{E} + \frac{d\mathbf{B}}{dt} = 0 \qquad (6.59)$$

where, $\mathbf{B} = \mu_0 \mu_r \mathbf{H}$ gives the $\mathbf{B}$–$\mathbf{H}$ relationship

From Ampere's law

$$\nabla \times \mathbf{H} = \mathbf{J} \qquad (6.60)$$

The transition of a superconductor from the superconducting state to the normal state can be estimated using Equation (6.61)

$$\mathbf{E} = \frac{E_0 |\mathbf{J}|^n}{J_c^n (B, \theta)} \qquad (6.61)$$

where, $E_c = 1 * 10^{-4}$ V/m and $n = 21$ are considered.

$J_c(B,\theta)$ at 77 K can be calculated using $I_c(B,\theta)$ from Equation (6.62)

$$I_c(B,\theta) = \frac{K_0}{\left(\dfrac{B}{\beta_0} + 1\right)^{a_0}} + \frac{K_1}{\left(\dfrac{B}{\beta_1} + 1\right)^{a_1}} \left[\left(\frac{B}{\beta_w} + 1\right)^{6/5} \cos^2(\theta - \theta_0) + \sin^2(\theta - \theta_0)\right]^{-0.5}$$

$$(6.62)$$

where, K_0, K_1, β_0, β_1, β_w, a_0, a_1 and θ_0 are the constants that depend on the model analysis.

AC losses can be estimated using Equation (6.63)

$$Q_{AC} = \frac{\displaystyle\int_{B_{min}}^{B_{max}} \left(\int_S \mathbf{E} \cdot \mathbf{J} dS\right) dt}{T} \qquad (6.63)$$

where, T is time period of each AC cycle and dS is cross-sectional area

The Jiles-Atherson hysteresis model is also used for estimating the magnetization effects in the inductive SFCL [35]

$$M = M_{irr} + M_{rev} \qquad (6.64)$$

where, M is magnetization, M_{irr} is irreversible magnetization, M_{rev} is reversible magnetization, $M_{rev} = c(M_{an} - M_{irr})$ and c is domain flexing parameter

The rate of change of irreversible magnetization is calculated using Equation (6.65)

$$\frac{dM_{irr}}{dH} = \frac{M_{an} - M_{irr}}{\lambda\delta - \alpha(M_{an} - M_{irr})} \tag{6.65}$$

where, δ is directional parameter, $\delta = \sin\left(\dfrac{dH}{dt}\right)$, λ is pinning site density parameter. If $\dfrac{dH}{dt} > 0$, $\delta = +1$ and if $\dfrac{dH}{dt} < 0$, $\delta = -1$ and M_{an} is Anhysteretic magnetization

From the basics of Anhysteretic magnetization, the relationship between the magnetic flux density (B) and magnetic field intensity (H) is given in Equation (6.66)

$$B = \mu H = \mu_0 \mu_r H = \mu_0 (H + M) \tag{6.66}$$

where, M is magnetic field intensity

The effective field intensity can be calculated as follows

$$H_e = H + \alpha M \tag{6.67}$$

where, α is scaling coefficient

The Langevin equation is used for phenomenological representation of Anhysteretic magnetization

$$M_{an} = M_{sat}\left(\coth\left(\frac{H_e}{a}\right) - \frac{a}{H_e}\right) \tag{6.68}$$

where, M_{sat} is saturation level

$$\frac{dM}{dH} = \frac{(1-c)\dfrac{M_{an} - M}{\sin\left(\dfrac{dH}{dt}\right)\lambda(1-c) - \alpha(M_{an} - M)} + c\dfrac{dM_{an}}{dH_e}}{(1-\alpha c)\dfrac{dM_{an}}{dH_e}} \tag{6.69}$$

The application of the Jiles–Atherson model to the two-coil-inductive SFCL gives

$$\frac{di}{dt} = \frac{V_0(t) - iR}{L\left(2 + \dfrac{dM_1}{dH_1} + \dfrac{dM_2}{dH_2}\right)} \tag{6.70}$$

where, R is load resistance, $V_0(t)$ is AC voltage source, L is inductance in the coil, $L = \dfrac{\mu_0 w_c^2 A}{l}$, w_c is number of turns in the coil windings and l is mean magnetic path length.

The uncoupling time of the SFCL [36] can be calculated using Equation (6.71)

$$t = \mu_0 \mu_r \frac{w^2}{\rho \pi^2} \tag{6.71}$$

where, μ_0 is permeability in free space, μ_r is permeability, w is width

The recovery current in the SFCL can be calculated using the Stekly criterion

$$I_r = \sqrt{\frac{W_\lambda PS}{\rho}} \tag{6.72}$$

where, W_λ is critical thermal flux, $W_\lambda = 0.5 *10^4$ W/m^2 for a horizontal conductor and P is perimeter of the superconductor

The critical current that an SFCL can withstand is calculated using Equation (6.73)

$$I_c = \frac{I_r}{\sqrt{\dfrac{2W_\lambda}{\rho D \left(J_{c,ov}^2\right)}}} \tag{6.73}$$

where, $J_{c,ov}$ is overall critical current density and D is diameter of the superconductor

The adiabatic longitudinal velocity propagation can be calculated using Equation (6.74)

$$v_{ad} = \frac{J}{C_p} \sqrt{\frac{\rho k}{T_s T_b}} \tag{6.74}$$

where, J is current density, C_p is volumetric specific heat, T_s is interface temperature, T_b is cryogenic bath temperature and k is thermal conductivity

$$T_s = T_c - 0.5 \frac{J}{J_c}(T_c - T_b) \tag{6.75}$$

where, T_c is critical temperature and J_c is critical current density of superconductor.

6.10 Issues Related to Cryogenics in SFCL

If DC currents are used to operate the superconductor, the losses encountered in the superconductor are zero. However, the time varying currents or fields in the superconductor result in hysteretic losses. The performance efficiency and reliability of an SFCL integrated in a smart grid intrinsically depends on the efficient cooling of the cryostat, to retain the superconductivity of the super-conductors wound to the SFCL coils during fault compensation for unmanned long-standing safe operation. The cryostat encounters sudden thermal loads from the unexpected fault currents and should withstand such load occur-rences for efficient operation. The superconductor can be operated at critical temperatures ranging from 4 K to 80 K to maintain the low temperature. In bath type cryostat SFCL systems, cryocoolants, such as Liquid nitrogen (77 K) and sub-cooled nitrogen ($\approx$65 K), are used for cooling. Hence, there is a neces-sity for refrigeration systems to absorb the heat from the cryogenic environ-ment and reject the heat to the surroundings. Refrigeration systems of the recuperative type include Brayton, Joule Thomson and Claude cycles and that of the regenerative type include Gifford–McMahon (GM), Stirling and pulse tube cycles. However, **Gifford–McMahon (GM)** cooler-based condensers are majorly used to cool the closed loop SFCL with negligible boil-off conditions. Sub-cooled liquid nitrogen can be used in closed SFCL systems where larger current densities are transported through the superconductor because it can absorb huge amounts of heat dissipated due to Joule heating during the fault. Further, sub-cooled liquid nitrogen maintained at a pressure above 1 atm can suppress bubble formation in the SFCL cryostat and hence, the thermal perfor-mance, electrical insulation, fast recovery and spatial uniformity in tempera-ture of SFCL can be enhanced during the fault current compensation. During the fault operation, the SFCL coils quench and the resistance of the supercon-ductor increases thereby resulting in Joule heating; the LN_2 or sub-cooled liq-uid nitrogen encounters enormous amounts of heat and the volume of gaseous nitrogen increases in the SFCL cryostat. Hence, thermal stability, reliability and protection for designing the LN_2-cooled SFCL systems become the major con-cern. The commercialization of SFCL is possible only with the stable operation by maintaining the temperature below the boiling point of LN_2 for efficient per-formance. In normal operation, sub-cooled liquid nitrogen is circulated with an operating pressure of 300 kPa at a temperature of 77 K because the boiling temperature is 87.7 K. The heat loads in the cryostat are removed using a single-stage GM cryocooler to maintain the pool at sub-cooled temperatures. Further, to pressurize the liquid nitrogen, Helium is circulated as a non-condensable gas into the cryocooler. For cooling the SFCL system, solid nitrogen (SN_2) with a critical temperature below 63 K can be employed as an alternative.

References

1. R. S. Dondapati, A. Kumar, G. R. Kumar, P. R. Usurumarti, and S. Dondapati, "Superconducting magnetic energy storage (SMES) devices integrated with resistive type superconducting fault current limiter (SFCL) for fast recovery time," *J. Energy Storage*, vol. 13, pp. 287–295, 2017.

2. S. Kar, S. Kulkarni, S. Sarangi, and V. Rao, "Conceptual design of a 440 V/800 a resistive-type superconducting fault current limiter based on high Tc coated conductors," *IEEE Trans. Appl. Supercond.*, vol. 22, no. 5, 2012.

3. G. M. B. Steven M. Blair, Campbell D. Booth, "Current-time characteristics of resistive superconducting fault current limiters," *IEEE Trans. Appl. Supercond.*, vol. 22, no. 2, pp. 1–5, 2012.

4. W. T. B. De Sousa, A. Polasek, R. Dias, C. F. T. Matt, and R. de Andrade Jr., "Thermal – electrical analogy for simulations of superconducting fault current limiters," *Cryogenics (Guildf)*., vol. 62, pp. 97–109, 2014.

5. A. Hobl, S. Krämer, S. Elschner, C. Jänke, J. Bock, and J. Schramm, "Paper 0296 – A new tool for the 'grid of the future,'" CIRED Workshop 2012, Lisbon, Portugal, no. 296, pp. 1–4, 2012.

6. W. Paul and M. Chen, "Superconducting control for surge currents," *IEEE Spectr.*, vol. 35, no. May, pp. 49–54, 1998.

7. S. Dutta, "Modelling and analysis of resistive superconducting fault current limiter," Proceedings of the 2014 IEEE Students' Technology Symposium, 28 February–2 March 2014, Kharagpur, India, 2014.

8. T. Yazawa *et al.*, "Superconducting fault current limiter using high-resistive YBCO tapes," *Phys. C Supercond. Appl.*, vol. 468, pp. 2046–2049, 2008.

9. Z.-C. Zou, X.-Y. Chen, C.-S. Li, X.-Y. Xiao, and Y. Zhang, "Conceptual design and evaluation of a resistive-type SFCL for efficient fault ride through in a DFIG," *IEEE Trans. Appl. Supercond.*, vol. 26, no. 1, 5600209, 2016.

10. S. Kar, R. Kumar, P. Konduru, and T. S. Datta, "Development of joints between two parallel SS-laminated BSCCO/Ag HTS tapes," *IEEE Trans. Appl. Supercond.*, vol. 20, no. 1, pp. 42–46, 2010.

11. K. Koyanagi, T. Yazawa, M. Takahashi, M. Ono, and M. Urata, "Design and test results of a fault current limiter coil wound with stacked YBCO tapes," *IEEE Trans. Appl. Supercond.*, vol. 18, no. 2, pp. 676–679, 2008.

12. E. A. Young, C. M. Friend, and Y. Yang, "Quench characteristics of a stabilizer-Free 2G HTS conductor," *IEEE Trans. Appl. Supercond.*, vol. 19, no. 3, pp. 2500–2503, 2009.

13. T. A. Coombs, K. Tadinada, R. Weller, and A. M. Campbell, "Predicting the effects of inhomogeneities in resistive superconducting fault current limiters," *Phys. C*, vol. 376, pp. 1602–1605, 2002.

14. X. Wang, H. Ueda, A. Ishiyama, M. Ohya, H. Yumura, and N. Fujiwara, "Numerical simulation on fault current condition in 66 kV class RE-123 super-conducting cable," *Phys. C Supercond. Appl.*, vol. 470, no. 20, pp. 1580–1583, 2010.

15. F. Roy, B. Dutoit, F. Grilli, and F. Sirois, "Magneto-thermal modeling of second-generation HTS for resistive fault current limiter design purposes," *IEEE Trans. Appl. Supercond.*, vol. 18, no. 1, pp. 29–35, 2008.

16. M. C. Ahn *et al.*, "Basic design of 22.9kV/630A resistive superconducting fault current limiting coil using YBCO coated conductor," *Phys. C Supercond.*, vol. 463–465, pp. 1176–1180, 2007.

17. P. Tixador and A. Badel, "Superconducting fault current limiter optimized design," *Phys. C Supercond. Appl.*, vol. 518, pp. 130–133, 2015.

18. V. Sokolovsky and V. Meerovich, "Analytical approximation for AC losses in thin power-law superconductors," *Supercond. Sci. Technol.*, vol. 20, pp. 875–879, 2007.

19. M. C. Ahn, D. K. Park, S. E. Yang, and T. K. Ko, "Impedance characteristics of non-inductive coil wound with two kinds of HTS wire in parallel," *IEEE Trans. Appl. Supercond.*, vol. 18, no. 2, pp. 640–643, 2008.
20. S. Sugita and H. Ohsaki, "FEM analysis of resistive superconducting fault current limiter using superconducting thin films by current vector potential method," *Phys. C Supercond.*, vol. 378–381, pp. 1196–1201, 2002.
21. W. Tiago, B. De Sousa, T. Mariano, L. Assis, and S. Member, "Simulation of a superconducting fault current limiter : A case study in the Brazilian power system with possible recovery under load," *IEEE Trans. Appl. Supercond.*, vol. 26, no. 2, pp. 1–8, 2016.
22. X. Zhang, H. S. Ruiz, Z. Zhong, and T. A. Coombs, "Implementation of resistive type superconducting fault current limiters in electrical grids: Performance analysis and measuring of optimal locations," *Supercond. Sci. Techn*, pp. 1–15, 2016.
23. J. Langston, M. Steurer, S. Woodruff, T. Baldwin, and J. Tang, "A generic real-time computer simulation model for superconducting fault current limiters and its application in system protection studies," *IEEE Trans. Appl. Supercond.*, vol. 15, no. 2, pp. 2090–2093, 2005.
24. G. T. Son, H. J. Lee, S. Y. Lee, and J. W. Park, "A study on the direct stability analysis of multi-machine power system with resistive SFCL," *IEEE Trans. Appl. Supercond.*, vol. 22, no. 3, pp. 3–6, 2012.
25. A. R. Devi and J. N. Kumar, "Simulation of resistive super conducting fault current limiter and its performance analysis in three phase systems," *Int. J. Eng. Res. Technol.*, vol. 2, no. 11, pp. 411–415, 2013.
26. N. Hayakawa, S. Chigusa, N. Kashima, S. Nagaya, and H. Okubo, "Feasibility study on superconducting fault current limiting transformer (SFCLT)," Cryogenics, vol. 40, pp. 325–331, 2000.
27. G. Didier and J. Lévêque, "Influence of fault type on the optimal location of superconducting fault current limiter in electrical power grid," *Int. J. Electr. Power Energy Syst.*, vol. 56, pp. 279–285, 2014.
28. S. Dutta and B. C. Babu, "Modelling and analysis of resistive superconducting fault current limiter," in *Proceedings of the 2014 IEEE Students' Technology Symposium*, 28 February–2 March 2014, Kharagpur, India, 2014, pp. 362–366.
29. Y. J. K. Y. J. Kim *et al.*, "Analytical design method of high-Tc coated conductor for a resistive superconducting fault current limiter using finite element method," *IEEE Trans. Appl. Supercond.*, vol. 20, no. 3, pp. 1172–1176, 2010.
30. D. K. Park *et al.*, "Experimental and numerical analysis of high resistive coated conductor for conceptual design of fault current limiter," *Cryogenics (Guildf).*, vol. 49, no. 6, pp. 249–253, 2009.
31. C. Kurupakorn *et al.*, "Simulation of electrical and thermal behavior of high temperature superconducting fault current limiting transformer (HTc-SFCLT)," *J. Phys. Conf. Ser.*, vol. 43, pp. 950–953, 2006.
32. S. Elschner, F. Breuer, A. Wolf, M. Noe, L. Cowey, and J. Bock, "Characterization of BSCCO 2212 bulk material for resistive current limiters," *IEEE Trans. Appl. Supercond.*, vol. 11, no. I, pp. 2507–2510, 2001.
33. M. Noe, *Superconducting Fault Current Limiters*. Woodhead Publishing Limited, Cambridge, UK, 2013.
34. Y. Jia, M. D. Ainslie, D. Hu, and J. Yuan, "Numerical simulation and analysis of a saturated-core-type superconducting fault current limiter," *IEEE Trans. Appl. Supercond.*, vol. 27, no. 4, pp. 1–5, 2017.

35. D. Sarkar, D. Roy, A. B. Choudhury, and S. Yamada, "Harmonic analysis of a saturated iron-core superconducting fault current limiter using Jiles-Atherton hysteresis model," *Model. Meas. Control A*, vol. 89, no. 1, pp. 101–117, 2016.
36. P. Tixador, Y. Brunet, J. Leveque, and V. Pham, "Hybrid superconducting a.c. fault current limiter principle and previous studies," *IEEE Trans. Magn.*, vol. 28, no. 1, pp. 446–449, 1992.

Index